Environmental Perspectives and Issues

Environmental Perspectives and Issues

Edited by
Dr. Suhaib A. Bandh
Dr. Javid A. Parray

CALLISTO
REFERENCE

www.callistoreference.com

Environmental Perspectives and Issues
Edited by Dr. Suhaib A. Bandh and Dr. Javid A. Parray
ISBN: 978-1-64116-037-7 (Hardback)

Callisto Reference,
118-35 Queens Blvd., Suite 400,
Forest Hills, NY 11375, USA

Visit us on the World Wide Web at:
www.callistoreference.com

Table of Contents

Preface

We live in the age and era of science and technology. Our life has indeed been dominated by the scientific inventions and technological innovations. This world would be an absolutely poorer place without science and technology. Our charming and worth living life itself would become absolutely bleak and boring in absence of the scientific inventions and technological advancement being witnessed by us all over the globe. As a matter of fact scientific inventions, discoveries and innovations have indeed ushered in a cataclysmic revolution all over the world and in almost every field of life. Volumes may be spoken and written to highlight the universally acknowledged importance and usefulness of science and technology. But at the same time we and our environment are confronting a number of challenges for our sustenance and as such we liked to write this book very briefly so as to generate awareness among the people for the protection of our environment. The book is systematically divided into chapters in such a way that it deals with major components of Environmental Science and the chapters offer an overall insight into various aspects of environmental science. This more concise presentation focuses on key principles, scientific methods and ideas, and life-long learning skills of students. It provides a solid foundation in scientific approaches to environmental problems and solutions. We have also tried to integrate information from a wide range of disciplines from both natural and social sciences to make it suitable for the non-science background students.

Finally, I would like to thank the entire team involved in the inception of this book for their valuable time and contribution. This book would not have been possible without their efforts. I would also like to thank my friends and family for their constant support.

Editor

Population Ecology

The term population has been derived from a Latin word *"populus"* meaning people. It may be defined as a group of organisms of the same species occupying a particular space at a specific time. A population is a self-regulating system that maintains stability in an ecosystem. The ultimate constituents of the population are individual organisms that can potentially interbreed. The individuals of a population are there either as a result of reproduction or they are actively transported by some phenomena like emigration or immigration or by passive movements by agencies like wind and water. Under favorable conditions a group of individuals increases in number. However, as the environment keeps changing, it acts as a natural check on population.

Some ecologists recognized following two types of populations:

1) Monospecific population: It is the group of individuals of only one species. Population is the term often used by ecologists for monospecific populations.

2) Polyspecific population: It is the group of individuals of more than one species. It is also referred to be as mixed population and ecologists often use the term community for polyspecific populations.

The population has various group characteristics which are as follows:

1) Population size

2) Population density

3) Populations dispersion

4) Natality

5) Mortality

6) Age structure

7) Life tables

8) Biotic potential

1) Population size: The size of a population is generally expressed as the number of individuals in it. The size of a population (N) at any given place is determined by the processes of birth (B), death (D), new arrivals from outside or immigration (I) and going out or emigration (E).

2) Population density: It is defined as the number of individuals per unit area or per unit volume of environment or it is defined as the size of a population in relation to a definite unit area. Density may be numerical density (number of individuals per unit area

or volume) or biomass density (biomass per unit area or volume).

Crude density: It is the number or biomass of individuals per unit of total space.

Ecological density: It is the number or biomass of individuals per unit of habitable space

i.e. available area or volume that can actually be colonized by a population e.g. if a plant species (as cassia tora) is found crowded in shady patches and few in exposed parts of a specific area. The density calculated in total area (shady and exposed) is called as crude density, whereas the density calculated in shady patches only is called as ecological or specific density.

The density of organisms in an area varies with season's weather conditions, food supply and many other factors. However, the number of organisms in a specific area will be determined by the size of the organisms and the tropic level it occupies. Generally smaller the organisms, greater is the abundance per unit area. Furthermore, lower the tropic level, greater is the density of the organisms.

3) Population dispersions: It is the spatial distribution of individuals or distribution of individuals relative to one another in a population. In nature, due to various biotic and abiotic influences, the following three basic population dispersions are observed:

Regular/ Uniform dispersion: It is the even spacing of individuals that occurs by chance in nature but is common in managed ecosystems such as croplands and orchards. It results from the intra-specific competition among numbers of a population.

Random dispersion: When the position of each individual is independent of the other, it is random dispersion. It can occur in those areas where the environment is uniform and resources are freely available throughout the year as in an evergreen forest.

Clumped dispersion: It is also called as clustered dispersion and is the most common type. The patterns of distribution are shown by the responses of the individuals to environmental changes, reproductive patterns, habitat differences, daily and seasonal weather changes and social behaviors e.g. the distribution of human beings is clumped because of social behaviors, economies and the geography of an area.

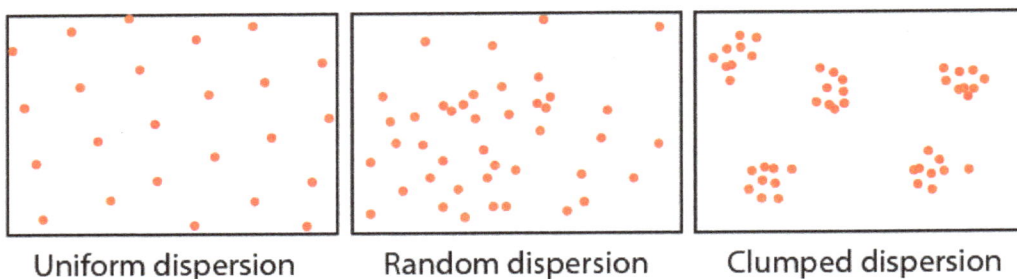

Uniform dispersion   Random dispersion   Clumped dispersion

4) Natality: Natality is a broader term used for the production of new individuals by birth, hatching, germination or fission. The natality rate is the production of new individuals per unit time. In case of human population, the natality rate is equivalent to birth rate. It is having a positive impact on the size of population. Natality is of two types:

i) Maximum Natality: It is also called as absolute, potential or physiological natality and is the theoretical maximum production of new individuals under ideal conditions (with no ecological limiting factors). It is constant for a population and is also called as fecundity rate.

ii) Ecological Natality: It is also called as realized natality or simply natality and is the production of new individuals under actual or existing conditions. It takes into account all possible environmental conditions and is also called as fertility rate. It is variable as it varies with size and composition of populations and with environmental conditions.

5) Mortality: It means the rate of death of individuals in a population or it is the probability of dying of individuals. The mortality rate or death rate is the number of deaths per 1000 persons in a population per unit time interval. It is a negative factor for population growth. Like natality, it is also of two types:

i) Minimum mortality: It is also called as specific or potential mortality. It represents the theoretical minimum loss under ideal and non-limiting conditions. It may be constant for a population.

ii) Ecological mortality: It is also called realized mortality and is the actual loss of the individuals of a population under a given/existing set of environmental conditions. As the environmental factors are variable so, the ecological mortality is not constant for a population. It varies with the variation in the environmental factors as predation, diseases and other ecological hazards.

For a population the surviving individuals are far more significant than the dead ones, the survival rates are of much greater interest than death rates. The survival rates are generally expressed by *Survivorship Curves.*

Survivorship curves: The pattern of mortality with age is best illustrated by survivorship a curve which is obtained by plotting the number of individuals surviving against the age. Three types survivorship curves are possible on the basis of survivorship and age and are presented as follows:

i) Highly convex curve (Curve A): This is the curve shown by those organisms which tend to live out their life span i.e. in which the mortality rate is low until the end of their life span. In these organisms the mortality rate is low in younger and adult age groups. The curve shows a slight dip in the early stages of the life span representing the death of a few individuals in the younger age groups. It is shown by large vertebrates including man.

ii) Highly concave curve (curve C): This type of curve is a characteristic of such species where mortality rate is high during the younger stages of life span. Some birds, fishes, short lived weedy annuals and many invertebrates exhibit such type of curve.

iii) Diagonal curves (curves B, B1, B2): This type of curve shows an age specific constant survivorship i.e. a constant mortality rate at every age. Some animals such as hydra, gull, American robins etc exhibit this type of curve. In fact no population in the real world shows a perfectly diagonal shaped curve (Curve-B). However, curves like B1 and B2 sigmoid curves are shown by many organisms. The shape of the survivorship curve may vary with the density of the population.

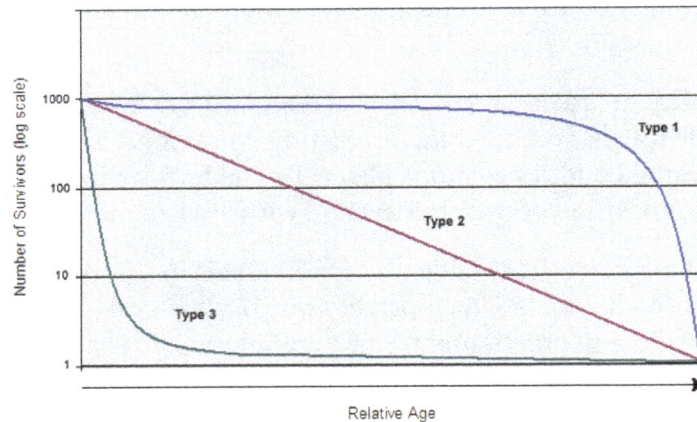

6) Age structure: In most types of populations, the individuals are of different age groups. The proportion of individuals in each age group is called as age structure or age distribution. From an ecological point of view there are three major functional or ecological ages (age groups) in any population. These are:

i) Pre-reproductive (or juvenile or dependent phase)

ii) Reproductive (or adult phase)

iii) Post-reproductive (or old age phase).

The relative duration of these age groups varies with different types of organisms. In humans the three stages are relatively equal in length about a third of their life falling in each class. This is an important characteristic of the populations as it influences both the natality as well as mortality of the population. The proportion of individuals of a population falling in pre-reproductive, reproductive and post-reproductive age groups is also referred to as the Age structures.

7) Age pyramids: It is the graphical representation of the proportion of different age groups in a population of any organism. It is also called as age-sex pyramid (OR) it is a vertical bar graph in which the number or proportion of individual in various age groups at any given time is shown from the youngest at the bottom of the graph to the oldest at the top. According to Bodenheimer (1938) there are three basic types of age-sex pyramids.

i) Pyramids with broad base (or triangular structure): It indicates a rapidly expanding population with a high percentage of young individuals and only few old individuals. Thus in rapidly growing young population, birth rate is high and the population growth may be exponential as in yeast, housefly and paramecium etc.

ii) Bell-shaped pyramid: It indicates a stationary population having an equal number of young and middle aged individuals. As the rate of growth becomes slow and stable the pre-reproductive and the reproductive age groups became more or less equal in size, post reproductive remaining as the smallest.

iii) Urn-shaped pyramid: It indicates a declining population having a low percentage of young individuals. This type of structure is obtained when the birth rate decreases drastically and the pre-reproductive age group dwindles in proportion to the other two age groups.

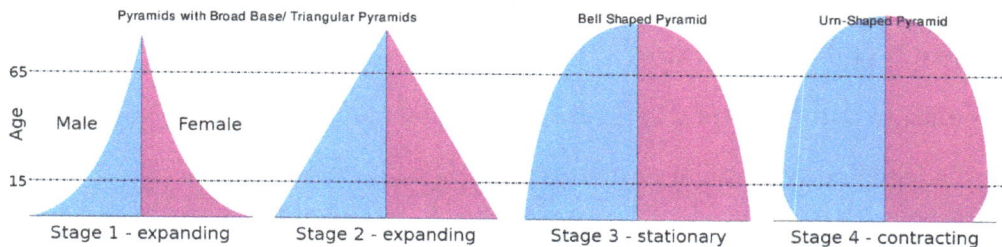

7) Life Tables: It is a statistical table which gives the vital information related to the average probability of survival or death at different ages, remaining life expectancy and the proportion of original births still alive. It consists of a series of columns, each of which describes an aspect of mortality statistics for the members of a population according to age.

The life tables are useful for computing the average longevity of a population, for showing the age composition of a population, for indicating critical stages in the life cycle (at which rate of mortality is high), for showing the differences between species and for showing the success of the same species in different biotopes.

8) Biotic Potential: The term biotic potential was coined by Chapman in 1928. He defined it as, *"the inherent power of an organism to reproduce to survive i.e. to increase in number. It is a sort of algebraic sum of number of young ones produced at each reproduction, the number of reproduction in a given period of time, the sex ratio and their general ability to survive under given physical conditions"*. Thus with the term biotic potential or reproductive potential we can put together natality, mortality and age distribution

r and k selection: The r and k classification was originally proposed by the two biologists Mc Arthur and E.O Wilson in 1967, who were interested in finding out the abilities of species to colonize different islands. The r and k refer to the parameters of population growth equations, where "r" is biotic potential and "k" is the carrying capacity.

r selected species: These are the species which tend to "boom" when environmental conditions are favorable and "best" when these conditions deteriorate. Besides they are having:

1) High rates of population growth

2) Poor competitive abilities

3) Good dispersal powers and are smelparous

4) Show highly concave type of survivorship curves

5) Small sized body and a brief life span.

6) Allocate more energy to reproduction and less to growth, maintenance and adjustment to the environment.

7) Live in harsh and unpredictable environments.

k selected species: These are the species which have relatively constant density at or near the carrying capacity (K) of the environment. They have:

1) Low rates of population growth.

2) Often exist in natural habitats and are hence good competitors

3) Poor dispersal powers and are itereoparous

4) Show highly convex type of survivorship curves

5) Large body size and a prolonged life span

6) Allocate more energy to non-reproductive activities

7) Live in stable and predictable environments.

Correlation of r and k selection

Life History Features	r-selection	k-selection
Development	Rapid	Slow
Reproductive rate	High	Low
Reproductive Age	Early	Late
Body Size	Small	Large
Life Span	Small	Large
Competitive ability	Weak	Strong
Population Size	Variable	Constant
Dispersal ability	Good	Poor
Survivorship	High mortality in young (concave curve)	Low mortality in young (convex curve)
Reproductive strategy	Semelparity (single reproduction)	Intereoprity (repeated reproduction)
Type of Habitat occupied	Disturbed	Non-disturbed

Population Growth Rates: Population is a dynamic entity i.e. it keeps on changing its size either by increasing or decreasing over a period of time. The rate is always obtained by dividing the change by the time elapsed during the change. The rate at which population is changing is more important than its size and composition at any moment. The growth rate of a population is expressed as the number of individuals by which the population increases divided by the time elapsed during the change. It can also be defined as the change of population per unit time.

Suppose N = number of organisms in a population.

t = Time

Δ= Change in any entity

Then

$$\text{Growth Rate (r)} = \frac{\Delta N}{\Delta t}$$

i.e. average rate of change in the number of individuals per unit change in time

$$\text{Specific Growth Rate} = \frac{\Delta N}{N\Delta t}$$

$$\text{\% Growth Rate} = \frac{\Delta N}{N\Delta t} \times 100$$

Population Growth Forms: Populations have characteristics patterns of increase which are

called as population growth forms. Such growth forms represent the interaction of two important factors, the biotic potential and the environmental resistance/stress. The growth is the most

fundamental dynamic factor that a population displays. Different populations in the world show two different types of population growth forms.

i) Exponential Growth forms: When a small group of animals is introduced into a suitable unoccupied habitat, it would tend to expand geometrically/exponentially (i.e. in the ratio of 2, 4, 8, 16, 32, 64...............). It represents the rapid growth in density with time.

The exponential growth form is expressed as:

$$\frac{\Delta N}{Nt} = rN$$

OR

$$\frac{dN}{dt} = rN$$

OR

$$\frac{dN}{\Delta t} = (b-d)N$$

Where

b = birth rate,

d = death rate,

r = biotic potential,

N = population size,

t = time,

dN = rate of change in numbers,

dt = rate of change in time and

$$\frac{dN}{dt}$$ = rate of population increase

The exponential growth form gives a J-shaped growth curve when the density value is plotted

against time i.e. the population increases exponentially until it over shoots the ability of the environment to support it, then the population declines sharply or crashes to cease abruptly through the density dependant factors like starvation, diseases, overcrowding and emigration.

ii) Logistic growth form: In 1938 a Belgium mathematician P.F. Verhulst provided a concept of decline in population growth as the density of population increases by rightly predicting that in the real world neither the environment is constant nor the

resources are unlimited. As the population density increases competition for the available resources among the members of the population increases leading to a decline in population density. Thus a more appropriate equation to explain this type of growth form is:

$$\frac{dN}{dt} = rN\frac{(k-N)}{k} -----\text{Logistic Equation}$$

Where

$$\frac{dN}{dt}$$ = rate of change in population,

k = carrying capacity (level beyond which no increase in population can occur),

r = biotic potential and

$$\frac{(k-N)}{k}$$ = proportion of resources unused.

The logistic growth form gives an S-shaped (sigmoid) growth curve when the density of the population is plotted against time i.e. when the individuals are introduced into an unoccupied area the growth of the population is at first slow (positive acceleration phase or lag phase), then becomes very rapid (log/logarithmic phase) and finally slows down as the environmental resistance increases (the negative acceleration phase) until an equilibrium is reached around which the population size fluctuates more or less irregular. This level beyond which no increase in population occurs is called as carrying capacity (k).

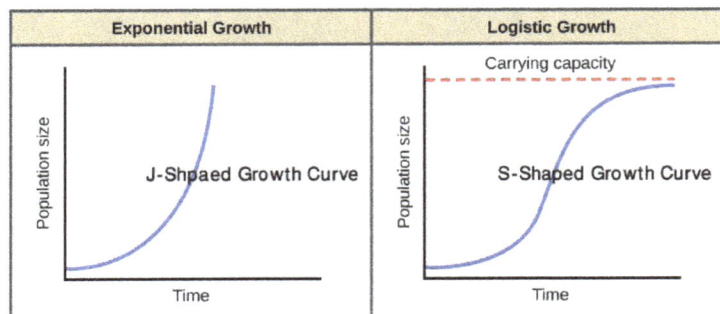

History of Human Population: Human beings evolved some 150,000 years ago on the planet earth. The humans of the Stone Age were not less than animals in their behavior, mode of living and food habits. In earlier times their population was very low by the interplay of various environmental factors like diseases and natural disasters. But the present day humans evolved around 250,000 years ago with a population of about 250 million in 1 AD rising subsequently to 545 million by 1650 AD. Historical records show that the human population doubled almost six times during a period of 13650 years meaning the average time required for each doubling was more than 2000 years. But the human population of the last century had doubled only in a period

of 40 years (from 2.5 billion to 5 billion). Before the end of the 20th century the human

population passed a mark of 6 billion and is estimated to reach 8-12 billion by the end of 21st century. This is because in every hour more than 10,000 new people enter the world at a rate of approximately 3/second. Of the existing more than 6 billion about half live in poverty and half in comparatively comfort. The factors affecting global population are fertility, mortality, initial population size and time. The stability in population requires a reduction in fertility globally and will take some time.

Theories of population growth: These are the theories which explain the levels of population in their economic, social and political environments. After studying the historical records, trends and events they provide the basis for equalization, judgments and policies. There are a number of population theories proposed by different demographers from time to time and of them a few are as follows:

i)   Malthusian Theory of populations: Thomas Robert Malthus born in England in 1766, a Professor of History and Economics, influenced by the writings of Benjamin Franklin and the Chinese Hung Liang Chi was very sensitive towards the deteriorating conditions of England during his times. He discovered that the potentialities of human population growth might surpass the resources required for its support. So, he gave his controversial population Theory. In 1798 he wrote *"An essay on the principles of population as it affects the future improvements in a society"* as the first edition and later wrote another edition to justify his view point on population.

Assumption of the theory: His theory is based on the following few assumptions

i)   Food is essential for the survival of human kind

ii)   Sex passion between the opposite sexes is natural and will remain so as ever

iii) There is a close relationship between the number of children and the living standard

iv) Agriculture would gradually function under the *"Law of diminishing utility"*

Salient features of the theory: In light of the assumptions Malthus stated in his theory that, *"population when unchecked increases in a geometrical ratio, while the measures of subsistence increase in an arithmetic ratio"* He declared that the population has a tendency to increase in a geometric progression (like 2, 4, 8, 16, 32, 64.............................) while the food production tends to increase in an arithmetic progression (like 1, 2, 3, 4, 5, 6, 7, 8.................................). The former is subjected to population checks as wars, diseases and famines and the later is subjected to the introduction of new developments in agriculture.

The factors like wars, famines, diseases, floods etc are the factors which temporarily strike a balance between the food production and population are called as POSITIVE CHECKS. He put forth another kind of check for population stability called as PREVENTIVE CHECK which includes:

- Late marriage (i.e. woman to remain unmarried till the age of 28 years) a moral restraint

- Modern methods of birth control (an artificial restraint)

Criticism of Malthus's Theory: The theory got a mixed response in the society. Some great thinkers and writers praised him for the theory and some criticized him. Prof. Grey expresses his view as follows *"it is safe to say that no respectable citizen has ever been so vilified and abused as Malthus"*.

Some important critical points raised about the theory are:

i) Prophecy about food increase is not true. He was not able to visualize the coming of a new scientific era which boosts the food production by scientific methods and by converting the barren lands into productive lands.

ii) Unnatural methods of population growth checking as forwarded by him were found to be against the good human behavior.

iii) There is no direct link between population growth and food increase at a particular place as explained by Malthus e.g. England produces only 1/6th of its total consumption but still the people there have sufficient to eat.

iv) Malthus failed to foresee the advancement in medical science.

v) Malthus took only one point of time into account, thereby failed to realize that time changes and the new ideas of new times boosts food production and other necessities of life and bring great favorable changes.

Demographic Transition Theory: It is the theory which throws lights on the birth rate and death rate and consequently the growth rate of populations, keeping in consideration the industrial revolution. It was propounded by W.S Thomson and F.W Notestine. According to E.G. Dolon the demographic transition refers to a population cycle which begins with a fall in death rate, continues with a phase of rapid population growth and concludes with a decline in the birth rate.

In underdeveloped countries the birth rates are high because of illiteracy, early marriage and demand for family labor and the death rates are also very high because of weak economy, health care facilities and standard of living, hence the growth rate is LOW- this is the old balance in population growth.

In the initial stage of economic development, the living standards rise, medical facilities enhance, health conditions improve and literacy rate increase which lead to fall in the death rates, but the birth rates are still higher, hence the population growth rate rises rapidly.

With the rise in economic growth both the death rates as well as the birth rates fall due to high literacy rates and higher living standards, resulting in a decreased level of population growth (zero level). This is the new balance in population growth. This shifting of population growth rates from old balance to new balance is known as demographic transition.

Stages of Demographic Transition: Some of the demographers like Thomson, Baiker and Hoover gave three general stages of demographic transition on the basis of death rates and birth rates.

i) Pre-transitional stage: this stage is characterized by:

- High birth (because of

    i) social beliefs and customs

    ii) Polygamy)

- High death rates (Due to

    i) Famines

    ii) Epidemics

    iii) Malnutrition

    iv) lack of health care facilities)

- Stationery growth rate

- No control stage

ii) Transitional stage: It is the stage of early development and is characterized by:

- High birth rate (because of already existing factors)

- Declining death rate (Due to

    i) Increased literacy

    ii) Health care facilities

    iii) Scientific development

- Rapid growth rate

- Population explosion stage

iii) Post-transitional stage: It is the stage of full development and is characterized by:

- Low birth rates (due to

    i) literacy

    ii) changed beliefs and customs)

- Low death rate (due to already existing factors)

- Stationery growth rate

- Zero population growth stage

Some of the Demographers divided the transition period into 4 stages

Stage 1: It is that stage of demographic transition in which both death rate and birth rate were high. During this stage there were no medical care facilities and standards of life were also low. Birth rates were high because of early marriage, illiteracy, religious beliefs and demands for family labor. However, the growth rates were low.

Stage 2: In this the modern health care facilities evolved, standard of life changed and finally began to drive the death rate down but birth rate remained the same which ultimately resulted in rapid population growth (population explosion).

Stage 3: In this stage the cost of supporting large families increased which discouraged parents to propagate large families. This resulted in the fall of birth rates, thus getting close to death rate. Stage 4: This is the last stage of demographic transition and is characterized by a higher but stable population size because both the death rates and birth rates are relatively low. The developed world remains in the 4th stage of demographic transition.

This demographic transition does not occur overnights but gets completed in decades together during which the population grows rapidly.

Factors regulating population size: Population regulation means maintaining the number of individuals of a population at a particular level usually the carrying capacity. In some of the populations the size is regulated by the extrinsic environmental factors where as some of the species are provided with unique and intrinsic self-regulatory mechanisms such as failure of reproduction and self-initiated mortality. Generally viewed the various factors that regulate the size of a population are as follows:

1. Competition: Competition for natural resources such as food, water and space results only when these resources are in short supply. As long as these resources are abundant, they allow the population to grow and reproduce without any competition. The population grows exponentially till the resource is sufficiently available but when the resource availability decreases; they compete with each other and with the result the fittest individuals survive and others perish.

2. Weather and climate: These are density independent factors and regulate the population size usually by influencing the availability of food. Changes in population growth can also be correlated directly with variations in moisture and temperature.

3. Territory: It is the area actively defended by an organism. There are some forces that cause the isolation of an organism from other and these individuals then actively defend these areas called territories. These initiating factors are various such as competition for resources, behavioral response and chemical isolation mechanisms. These territories are mostly made by organisms especially for nest building, egg laying and parental care for the young ones. As aggregation may increase the competition in these areas and hence regulate the population size. Territoriality might regulate population density, but only under certain

circumstances.

4. Predation: It is the consumption of one living organism by other, a relationship in which one organism is benefited and other one is harmed. It is called as prey predator relationship. Here when the prey population increases, the predators also increase. But when the population of predators increases the prey population once again decreases because of the increased frequency of prey attacks. Hence the population is regulated.

5. Natural disasters: Volcanic eruptions, earthquakes, landslides and floods cause sufficient damage to property and life. Large number of casualties occurs. Hence these events also regulate the size of a population.

Intra-specific competition: Competition occurring within the individuals of a species is referred as intra-specific competition. It involves the behavioral and physiological interactions among the members of a population. Increased density may affect population growth through physiological responses of individuals resulting in abnormal growth, behavior, degeneration and infertility. Social pressure and crowding may also induce immigration or dispersal. In general leading to the regulation of population size.

Human Population

The human population has already crossed the mark of 7 billion and is still growing. Every day about more than 25000 new individuals and every year about 90 million people are added to the earth. It is estimated that by 2025, 3 billion more people will be added to the earth. Of the total increase in population about 90% occurs in underdeveloped countries or developing countries, at a rate of 2% in most of them and 3% in some.

In India, as per the census of 2001 the population was 102.7 cores and is expected to reach 117.5 to 140 cores during 2010-2025. We alone account for about 16% of the world's population with a mere 2.4% of globe's land resource. If the rate of population growth continuous in the same fashion, we may overtake china by 2045. We have already crossed the mark of 1 billion by 2000 and are expected to reach unmatched figures in near future. According to 2001 census, the density of population is highest in west Bengal where it is 904/km2, then comes Bihar with 880/km2. The most thinly populated states are Arunachal Pradesh, Mizoram, Sikkim, J&K, Meghalaya and Himachal Pradesh.

Population Growth and Decline:

The Growth and decline are the two basic characteristics related to populations. There are three factors namely birth, death and migration, which make the populations to grow or to decline.

1. Births: These are measured by using birth rates. It is the member of new born babies per 1000 persons in a population in a given period of time. It is a factor responsible for the growth of a population and is called as the positive factor.

2. Deaths: These are measured by using death rates. This is the number of deaths per 1000 persons in a population per unit time (a given year). It is responsible for the decline of a population and hence is a negative factor. When the birth rate in a population is greater than death rate it grows and when birth rate is lower than death rate it declines. When the two are same then the population size remains constant and shows zero population growth.

3. Migration: It is the movement of a population from one area to another and then back again or it is the periodic departure and return of the individuals of a population. This also affects the population size in a particular area. It has two main forms:

    i) Emigration: It is the one-way outward movement of individuals under natural condition it occur at a particular place when overcrowding takes place there. It leads to the occupation of new areas elsewhere. Continues emigration is rare and when it occurs it result in depopulation.

    ii) Immigration: It is the one-way inward movement of individuals. Immigration leads to a rise in population level causing an overpopulation which lead to an increase beyond the carrying capacity.

Various other factors contributing towards the increase in population are:

- Decreased level of IMR (infant mortality rate)

- Increased life span due to better

  o Sanitation

  o Community health

  o Health care facilities and

  o Availability of life saving drugs etc.

- Increased agricultural productivity by scientific improvements

- Religious beliefs, traditions and cultural norms

- Non-availability of contraceptives to the populations living in remote areas.

Ill effects of over population: Huge population means huge problems. It proves to be a hindrance to economic development of the under developed countries. Main ill effects of over population are as follows:

i) Problem of food: In spite of green revolution the food taken by the Indians is poor in calories. Therefore, a vicious circle of poverty is set in motion.

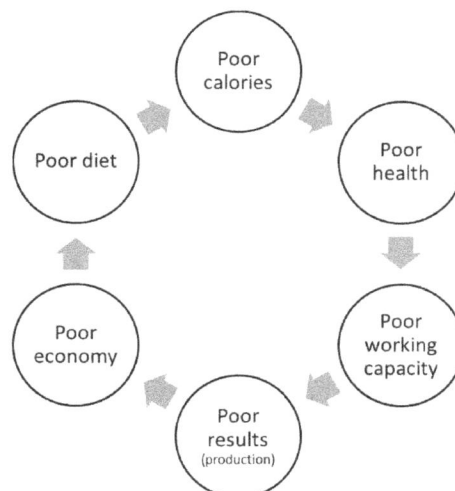

ii) Problem of shelter: Due to over population, millions live on pavements, procreate on pavements and die on pavements. If houses are built, both agricultural land as well as forest land will get reduced.

iii) Unemployment: Providing gainful employment to everyone becomes awfully difficult when there is lack of resources and over expanding population.

iv) Education and literacy: Less educational facilities. In India about 35 cores of population are still illiterate.

v) Effects per capita income: Since the growth rate of population is for high than growth rate of national income the per capital income begins to diminish/decline.

vi) Rate of inflation increases: Due to increasing gap between demand and supply-rates become variable and life is becoming costly.

vii) Big Ques everywhere: Crowding of trains, buses, shops, offices and all other public places leading to wastage of time and energy.

viii) Social overhead facilities: Over population necessitates social overhead facilities like transport, health, education, water supply etc. for the additional population. This leads to the shortage of resources for other developing projects and welfare activities.

Consequences of high population growth on environment (in Indian context):

The uncontrolled growth of population is causing many socioeconomic problems, thereby ultimately disturbing the overall ecological balance. The main consequences of population growth on the environmental segments in India are:

1) Here the population is growing alarmingly and needs more and more food production. It forces us to raise crops repeatedly and one after the other. For maintaining the fertility of the soil the increased doses of chemical fertilizers are used leading in the disturbance of soil composition. The unrestricted use of these chemicals is posing many dangers to the humans as well as the ecological balance.

2) The diminishing productivity of soil as a result of intensive agriculture has been increasing costs of production thus increasing the burden of subsides on the economy. This problem is further aggravated by the uneven distribution of poverty in the country.

3) India with the growing population has been turning forestlands into farmlands. Today the land under forests is less than 20% of the total area. But, for a balanced economy it should be at least 33%.

4) India is facing an intense crisis over the resources. There is intense competition for the nation's limited resources leading to aggressive disputes in the human societies. The limited resources are being over exploited because the living standards are rising. The rise in the standards of living means higher consumption of articles which in turn means consumption of resources on much faster rates.

5) There is unequal sharing of natural resources like water, mineral, forests, and energy etc. This leads to the over exploitation of the resources by the poor class illegally. Thus, leading to an ecological disturbance.

6) Rapidly growing population burdens the available infrastructure and opportunities. Thus leading to a situation of congestion of urban areas, putting an unbearable strain on urban amenities.

7) The rapid growth in population has adverse effects on the equitable distribution of income. The increase in gross national production (GNP) is greatly reduced due to it.

Need to Control Population Growth:

To break the vicious circle of poverty set into motion by the ever increasing human population we have to make earnest efforts at controlling the population, besides conserving our natural resources and the environment. To achieve the goals of development-like improvement of the income levels, educational facilities, clean drinking water facilities and sanitation etc. it is high time to go for the modern birth control practices and promotion of the small family norms. This could be achieved by the combination of development and contraception approaches. To accelerate the process of fertility control in the populous areas of the country through well planned and efficient family planning programme is very crucial at this juncture.

Family Planning:

In India it is the need of the hour. It is also called Small family norm or Family welfare programme. It was launched for the first time in the world by India in 1952 a nationwide programme, with the aim to decline the birth rates. The concept of family planning means *"enlarging the size of the family up to a certain limit in keeping with size of the income of the family"*. It implies *"to have children by choice and not by chance, by design and not by accident"*. Proper spacing of children and conscious acceptance of small family norms are the thrust areas of this programme. The family planning was a government sponsored programme but even then there was no marked decline in the birth rate. Then the government of India introduced the concept of Compulsory sterilization on 16th April 1976 (NPP) which was aimed at the sterilization of one of the parents after two children. But it faced a heavy contradiction from the general public. It then became a voluntary programme and came up with the slogan *"HUM DO HUMARE DO"*. Now-a-days this programme is more successful in the developed countries rather than in developing countries.

Reasons for its Failure in Developing Countries:

1) Poverty: Among the poor class the myth of *"more children more income"* is still prevailing.

2) Illiteracy: Due to wide spread illiteracy more of the people don't know the benefits of this programme.

3) Fatalists: They regard each child as a gift of God. They believe more on fate.

4) Religious opposition: Some regard it as anti-religious and an immoral act.

5) Lack of publicity: particularly in the rural areas.

6) Shortage of trained staff: There is shortage of trained staff including doctors and nurses.

7) Unmet needs of family planning: There is a great unfulfilled need of family planning in the world. In India alone there are about 29 million women with unmet needs of family planning.

8) Opposition from husband and in laws to go for family planning.

9) Fear about the side effects of contraceptives.

10) Lack of cheap and effective methods of birth control.

Methods of Family Planning:

1) Sterilization: it is done in two ways

   a) Male sterilization (vasectomy): it is an approach hardly criticized and resisted in India and accepted only by 3% people.

   b) Female sterilization (tubectomy): as per the national family health survey conducted in India it is most widely accepted and used by 67% people. This signifies that the programme is mainly female centered.

2) Use of IUD's (inter uterine devices): it is a cheap, simple and effective method.

3) The condom: it is very cheap and effective method.

4) The Pills: it is the most popular method of family planning but should be used with care because of their side effects e.g. saheli, unwanted 72 etc.

5) MTP method: it is medical termination of pregnancy and is commonly called as abortion. It is actually not involved in family planning methods. But was passed in 1971 in India and came into force in 1972. It is allowed only:

   • If the pregnancy involves a risk to the life of mother and baby.

   • If it is an illegal case.

   • If age of woman is below 18 years.

   • If the woman is not mentally and physically sound.

Education:

It can play an excellent role in reducing the population growth by making the individuals of reproductive age group aware about the

   • Advantages of small family norms

   • Ill effects of rapid population growth

   • Ill effects of large families and

   • Ill effects of population explosion.

The education can be provided through radio, TV, mass media, poster, pamphlets etc. For this the following educational programmes should be carried on:

1) Adult education: It is an educational programme which is aimed at the development of decision making skills in the individuals. One should be able to make the decision:

   • Regarding fertility

- Regarding the development of children

- Regarding the preparedness for motherhood

- Regarding the traditional beliefs of having children soon after marriage.

Women and girls are the key target groups of the programme and should be informed about the health issues and human rights.

2) Sex education: It is an educational programme which covers the anatomy and physiology of reproductive systems, physical, emotional and psychological changes during puberty, conception, pregnancy and birth. It also deals with the sexual behavior and sexually trans-mitted diseases (STD). It is aimed at providing knowledge about:

- Biological

- Psychological

- Socio-cultural and

- Moral dimensions of human sexuality

3) Family life education: It is the educational programme which was supported by UNESCO in 1970 and is closely related to sex education. It is an educational process designed to as-sist the young people in:

- Physical development

- Mental development

- Social development

- Cultural development

- Moral development and

- Emotional development

To prepare them for

- Marriage

- Parenthood and

- Adult hood

It also deals with the social relationships of the families and societies.

4) Population Education: It is an educational programme which provides for a study of popu-lation phenomena so as to enable the students to take rational decisions towards problems arising out of rapid population growth. It is aimed at helping individuals to understand the interrelationship of population size and development.

Objectives:

1) To establish the link between population and development

2) To realize how rapid growth of numbers disturbs our nature

3) To explain the huge gap between developed and developing countries

4) To understand that the family size is controllable

5) To change the attitude towards family size

6) To acquire information about economic development and family welfare.

Status of Women:

In India the status of women was and is still very low (almost negligible) in some of the societies. They were tortured by husbands and in laws and were denied the better facilities of living, hence leading them to:

- Low nutrition

- Low health status

- Low literacy rates

- Low participation in

  o   Legislative

  o   Executive &

  o   Judiciary

- Low participation in the economic and political activities

- Economic dependence

- Be dominated by patriarchal norms

- Gender oppressions

- Tortures for sex determination although it is biologically determined

- Inadequate attention to "Gender justice" and "woman Empowerment".

Of late the "Gender Justice" and "women Empowerment" received due attention due to:

1. UN conference on population held at Cairo in 1995

2. International conference on women (Beijng) 1995

3. Social Summit Conference (Copenhegan) 1995

4. UNDP's Human Development Report 1995-96

The GoI also laid emphasis on raising the status of women by:

1. Constitutional measures (Article 610A) – envisages that it is the constitutional duty of every citizen to renounce the practices derogatory to the dignity of women.

2. Including its objectives in the five-year plans since 1970's

3. The formulation of –National Commission for women and National policy on empowerment of women

4. Reserving seats for them in Panchayati Raj system.

Keeping in view the above mentioned facts all the nations of the world emphasized that the need of the hour is to:

- Give high status to women

- Give more weightage to "right to equality"

- Give them the better facilities of living (at par with men).

Thus leading them to:

- Better nutrition

- Better health status

- Access to educational facilities

- High literacy rates

- Equal participation in

  o Legislative

  o Executive &

  o Judiciary

- High participation in economic and political activities

- Economic independence

- Freedom to women

- No gender oppression

- Adequate attention to "Gender justice" and "Women empowerment"

- Equal opportunities in power sharing

This is the key to the rest of her future and development. The problem with our population control programme is that unless the status of women is improved there is no hope of curtailing it. Her status in the society is the key to the success of population control programme in India.

Economic Growth:

The relationship between population and development has become an issue to debate on. The economic growth may accelerate population growth in the beginning but the rapid increase of population may have an adverse effect on the economic growth because we will be left with limited resources to fulfill our needs.

In the developed countries the population growth is very rapid in the early stages of development but later declines with:

- Improved standards of living and

- Improved literacy rates.

India started planning for development in 1951 and was soon caught up with problems of rapid population growth. The relationship between population and development can be put up in a following few ways:

1. Development depends on population i.e. population is an independent variable and development is a dependent one. According to this thought population should be modified to suit the prevailing economic conditions. If development is not fast enough it would result in poverty and overpopulation, because it would not meet the growing needs of population. This means the developing countries need to check their populations, if they wish to develop.

2. Population depends on development i.e. development is an independent variable and population is a dependent one. This thought advocates that economic development helps in declining the population growth. Hence it can be concluded that an independent population policy could do no good to the rapidly increasing human population but a feasible policy on economic development can.

3. Both population and development affect each other and have a reciprocal cause and effect relationship. This is a more realistic view about the relationship of the two variables and has got the support from the world population plan of action. With the economic growth parents become concerned for the reduction of family size in order to provide adequate facilities to their children in future. Although the industrial growth has reduced the population growth but still a lot more is to be done.

3

Threats to Ecosystem

An ecosystem is a community of living organisms involved in physical, chemical and biological interactions between themselves and the non-living components (OR) An ecosystem is any ecological unit that includes all the organisms which interact among themselves and with the physical environment. Human beings and other living creatures' dependent on nature for their survival and existence as nature has been kind to them since their appearance on earth. Now on one side the human population is growing at a very fast rate and on the other side man has started utilizing the natural resources quickly and at a very large scale as compared to the pre-historic times. The combined effect of both the situations leads to a condition where non-renewable resources get exhausted at an alarming rate. The non-availability of resources leaves an adverse effect on the overall balance of the ecosystems. The factors which appear as a threat to the existence of ecosystems are as follows:

Habitual Destruction:

Habitat destruction is the main threat to many species and is caused mainly by massive clearance of forests, expansion agriculture, changing land use patterns, increased demand of land for developmental activities, draining or filling up of the wetlands for development e.g. any highway that runs along the river may prevent several animals and birds from reaching the water body; it will also increase the run off of water, sediments and pollutants into the river thus resulting in the destruction of these ecosystems.

Genetic Erosion:

The loss of genes and genetic diversity is called as genetic erosion. One of the threats to genetic diversity is the reliance of humans on certain species and excluding the other varieties. We are promoting a narrow genetic base, i.e. a few genes with the highly desirable characteristics to increase production of desirable products thus replacing the thousands of primitive varieties on earth.

Impounding of Water:

Water is impounded by the construction of dams, reservoirs etc. The process of construction results in shifting cultivation, deforestation, destruction of wild life (including the wild relatives of crops, trees and animals etc.). Thus loss of the genetic resources of some valuable species occurs here.

Loss of Biodiversity:

Biodiversity is the variety of the world's living species including their genetic diversity and the communities and ecosystem they form. Dense tropical forests, have served as habitats of an innumerable ecosystems possessing distinct and varied biodiversity which maintains the integrity of the ecosystem. Hence these ecosystems maintain the equilibrium of the nature and natural ecosystems.

The loss of biodiversity occurs by large scale cutting of forests, excessive harvesting of plants and animals, indiscriminate use of pesticides, draining and filling of wetlands, destructive fishing practices, air pollution, disturbance of land use pattern. All these activities are a result of the excessive growth in human population. All this happening in nature is threatening the worlds critical ecosystems.

Expanding Agriculture:

The practice through which specific crops are cared for and managed so as to obtain maximum yield of plants and their parts for consumption is called as agriculture. The practice is usually done on agricultural fields which are limited in size to feed the ever increasing human population. In order to meet the demands of food for the expanding populations, agriculture faces green revolution through industrialization, use of chemicals, conversion of forests and wetlands to agricultural land (from 4.55bha in 1996 to 4.93bha in 1999) and application of fertilizers. All these practices have disrupted the valuable and unique food chains in soils, fresh waters and marine water ecosystems.

Causes for Concern:

The large scale expansion and intensification of agriculture has raised concern about the state of the agro ecosystems, as surveys show that about 40% of the world's agricultural land is already under a serious threat of degradation. Hence there are growing concerns about the stresses imposed by agricultural practices which include:

- Soil erosion

- Depletion of nutrients from soil

- Soil salinization

- Water logging

- Reduction of genetic diversity of crops

Moreover, the amount of land fit for agriculture is continuously decreasing because of nonagricultural uses in developed countries. Such activities seriously affect the soil structure and its biota through reduction of organic matter. There is also loss of water infiltration, reduction of moisture content and soil organisms in it.

The increased influx of agricultural chemicals further deteriorate soil conditions by reducing the abundance and diversity of soil organisms leading to a decreased agricultural productivity as well as damaging the system functioning. It also leads to the pollution of water bodies and thus having far-reaching impacts on aquatic life. Besides it is interfering with the landscape complexity as well by eliminating the woodland, hedgerows, fallow fields and individual trees, resulting in the loss of habitats for wild flora, fauna and insects, including the valuable wild relatives of domesticated plants and animals.

Thus the wide spread impacts of agriculture on biodiversity by changing agricultural practices, technologies and land use practices are becoming a cause of concern. So, the need of the hour is to strike a balance between the agricultural production and biodiversity.

Wastes from Human Societies:

All the types of environmental damages have been an outcome of the efforts for the ever increasing human population. The environmental damages are further intensified by our consumption habits, technological developments and particular patterns of social organizations and resource management. The loss of biodiversity, global warming, ozone depletion, acid rains, deforestation, water shortage, energy crises and loss of top soil are the indicators of the environmental stresses. These problems are the outcome of the activities of both developed and developing countries.

In developing countries wood used as firewood is 82% and for industrial use is 18% whereas for developed countries the figures are 16% and 84% respectively. The developed countries account for worlds 23% population and 85% gross waste product, but account for the largest part of mineral and fossil fuel consumption resulting in significant environmental problems. The increased human population causes increased utilization of resources and hence increased waste production. The wastes produced from these human societies are as follows:

1) Sewage: It is the collection of waste water from all the drains in homes, in other buildings and flushing's from toilets. It includes human excrete, paper, cloths, detergents, oils and greases etc. These wastes are dumped into the water bodies, thereby reducing the self-purifying ability of water bodies thus forcing the aquatic life towards extinction. Water becomes unfit for use and there occurs a reduction in the water borne diversity by reduction of their reproductive ability. The oxygen level of these water bodies is depleted thereby destroying the aquatic life. The enrichment of the water by nutrients leads to more critical conditions called "Eutrophication" resulting in the growth of algal blooms and release of toxins in the water bodies.

2) Industrial wastes: All types of industries are an outcome of the human efforts. The wastes generated from these can then be included in the wastes from human societies.

   They include:

i) Non-process wastes-such as office wastes, packing wastes etc. which are common to all industries.

ii) Process wastes-which depend upon the type of the product being manufactured such as tannery wastes, dying wastes, food processing wastes, plastic wastes, rubber wastes, metal scraps etc. from the respective industrial establishments.

   Heat and radioactive wastes are also added to the wastes from power plants and nuclear power plants respectively. The wastes generated form these industries are having adverse effects on both flora and fauna.

3) Agricultural wastes: The wastes generated from farms, agricultural fields, feed lots, livestock units etc. are called as agricultural wastes. They include paddy, husk, bagasse from

sugar cane, tobacco and slaughter house wastages. Besides, they also include pesticides and synthetic fertilizers intensively used in today's agriculture. Due to the wide use of these pesticides in agriculture they enter the food chains and get accumulated in successive tropic level of the food chains causing biomagnifications which ultimately kills many species and the entire ecosystem.

4) Municipal wastes: These include garbage and rubbish from our houses, offices, hotels, markets and also the street refuse such as street sweepings, dirt, leaves, litter etc. The term "rubbish" is used to denote non participle or non-biodegradable solid wastes which include combustible materials (e.g. paper, clothes and plastics etc.) as well as noncombustible materials (e.g. broken crockery, metals, glass, masonry wastes, metal canes and containers etc.)

Conservation:

The excessive growth of human population resulted into expanding needs of man. With the growth of industry and technology, man started utilizing natural resources at a much larger scale and a faster rate. Continuous increase in population caused an alarming increase in the demand for resources; rather it outpaced the rate of their utilization. This has created a situation when the non-renewable resources may come to an end after some time. But there must be some sort of balance between the population growth and the resource utilization. Thus, there is an urgent need of resource conservation in the world. It may be defined as, *"The management for the benefit of all the life including humankind of the biosphere so that it may yield sustainable benefits to the present generation while maintaining its potential to meet the needs and aspirations of the future generations"* OR *"the wise and judicious use of the resource"*. Conservation seeks to prevent all the necessary components of human life from complete elimination. It is embracing the preservation, maintenance, regeneration, sustainable utilization, restoration and enhancement of natural environment.

Aims of Conservation:

1) Preservation of the quality of environment

2) Ensuring a continuous yields of material both living and non-living by establishing a balanced cycle of renewal and harvest

Objectives:

1) Maintenance of essential ecological processes and life support system

2) Preservation of biological diversity 3) Existence of species and sustainability of ecosystems

Approaches of Conservation:

There are two general approaches of conservation:

1) In-situ-conservation: in-situ conservation also called as on-site conservation is the conservation carried out within the natural habitats by extracting and eradicating the harmful

factors from the habitats. It is ensured in areas like national parks, sanctuaries, biosphere reserves, sacred groves, nature reserves etc.

2) Ex-situ-conservation: ex-situ conservation also known as off-site conservation is the conservation carried out within the areas outside the natural habitats under proper human care. In this the organism is assured of food, water, air, space, health care and other facilities. It is ensured in zoos, botanical gardens, gene banks, seed banks, and germplasm banks by the techniques of cryopreservation and tissue culture.

Critical Status of Indian Forests:

Forests are the important components of our environment and economy. They check air pollution, soil erosion, wind erosion and save the hill slopes form landslides. They regulate the temperature and rainfall patterns in an area. They are the important natural resources providing various types of goods and services. The major product supplied by forests is wood and the minor products include gums, resins, medicines, tannins, fibers, lac, and canes etc. Moreover, forests provide shelters to wild animals; maintain the gaseous balance of atmosphere, water holding capacity of soil and climate of various areas. The total dependence of man on these forests has led to a significant reduction in the worlds forest cover. According to an estimate the forest cover of the world was:

i)   7000 million hectares in 1990

ii) 2890 million hectares in 1975

iii) 2370 million hectares in 2000

The maximum reduction in forest cover has been in developing countries (where the reduction is reported to be 41-63%). These days India with a forest cover of 22.7% (the effective forest being only 15%) is among the poorest countries in terms of world's per capita availability of forestland which is 0.01 ha/person as compared to the world average of 1.1 ha/person. India's forest cover comprises only 0.05% of the world forest cover and is losing about 1.5 million ha every year and if the same trend continues we will reach the *zero forest level* with a few decades. This large scale deforestation has been attributed to the mushroom growth of developmental projects.

Conflicts Surrounding Forest Areas:

Different tribes with different cultural setups live either on the edges of forests or within the forests and depend wholly and solely on the forest products. In recent years they have experienced a difficulty in gaining access to these forests because of various reasons as Govt. regulations, population pressure, declaration of state forests and national park or wildlife sanctuaries. In many countries the plans of forest conservation have failed because of ignoring the needs and rights of these tribes. Various conflicts surrounding these forest areas are:

1) Tribals and their rights: More of the tribals live in forests and use their product for their survivals. They are closely associated with each other. The tribal are getting solely everything from the forest and are in turn protecting them. But over the past few years many tribal communities have broken the tribal discipline and eroded the tribal culture of protecting these forests but are turning out to be the active agents of forest destruction. This

is further intensified by the encroachment of the tribal lands in forest areas by the people living in nearby villages. All this lead to:

- Overexploitation of wildlife products

- Killing of animals and

- Disturbance of tribal life

The tribals are having the rights to get the products like fodder, timber, vegetables and fruits from the forest (but not on whole) with the duty to act as the conservators of these forests. The national forest policy of the country has now realized the need to associate tribals in the forest conservation by recruiting them as guards and watchmen. So, conservation is possible only when we involve the tribals in the process of conservation.

2) Developmental Projects-Dams: The establishment of the Developmental projects the part and parcel of a country's progress leaves millions of farmers and laborers in any country homeless and landless. They are having great environmental hazards associated with them some of which are as:

a) Impacts within and around the area of activity

b) Downstream effects caused by the alteration of water courses

c) Regional effects including resource utilization and socio economics aspects

The Impacts of Dam Construction Include:

- Changes in microclimate

- Loss of vegetation

- Soil erosion

- Variation in water table and

- Seismic activity etc.

Eminent scientists believe that the social, environmental and economic costs of these constructions is far more than their benefits by the mounting opposition from various scientists and environmentalists. The Govt. has been forced to review a number of Dams in the light of their impacts on tribal's, flora, fauna, vegetation (generally forest). Some of them which gained importance due to the efforts of people are as follows:

i) Narmada valley project: The Sardar Sarover Dam project, world's largest River valley project has drawn the attention of the whole world due to the activities of "Narmada Bachao Andolan". It consists of 30 big dams and more than 3000 medium and small dams. It is threatening the existence of several species of birds, animals, plants and microbes. It is estimated to displace over one million people, submerge 56000ha of agricultural land and 60,000ha of forest land, thus sparking worldwide controversy. It is expected to threaten the livelihood of more than 140000 people in the areas to be flooded by it.

ii) Tehri Dam: The big dams have always been a matter of controversy and The Tehri dam also has its share of dispute. The 260.5-meter-high earth and rock fill dam being constructed at Tehri on the river Bhagirathi on the foothills of Himalaya is expected to affect 21000ha of land, submerging of some 109 villages along with the Tehri Town. It has already displaced some 85000 people from the area. The dam is expected to impound large amounts of water, thus causing earth tremors and submergence of fertile agricultural land.

Scientific Forestry and its Limitations:

Forests can be wonderful, beautiful and productive places with sparkling, bright pools, dark swamps, high mountains, mature trees, fresh seedlings, ancient trees, big and small animals etc. needing a wide range of habitats. So, forestry should be having multiple dimensions rather than the mere production of biomass and tree cropping. Although science can help forests but scientific forestry is based on single objective. Its obstacles are as follows:

i)   It aims at clearance of ancient forest and wood which will have harmful impacts such as soil erosion, loss of habitat, loss of biodiversity and wild life etc.

ii) It takes very long time to reach maturity

iii) It emphasizes on wood production and neglects nature and soil conservation.

iv) It does not involve the communities in management and neglects their rights over these forests

Social Forestry: The social forestry programme (SFP) was initiated by GoI as a nationwide programme in 1976, by making 175 million acres of wasteland, deforested land, and average private land etc. available to it. It was initiated with the aim to provide forest-related needs to the rural communities. The total area planted between 1980 and 1989 was 11.8 mha, at an average of 1.3 mha /year till 1994. The national wasteland development board (NWDB) was set up as a coordinating agency and was involved in designing three general types of programmes as:

i)   Creation of strip plantation along roadsides

ii) Utilizing communal lands for mixed species planting

iii) Farm forestry or Agro forestry on private farm lands

Objectives:

• Fuel wood supply to rural people

• Small timber supply

• Fodder supply

• Protection of agricultural fields from erosion

The national commission on Agriculture defined social forestry to include farm forestry, extension forestry, reforestation in degraded forests and recreation forestry.

Farm Forestry: Aimed at growing trees on bunds and boundaries of farmers' fields and is taken up by farmers themselves.

Extension forestry: It is a mixed programme of growing trees on wastelands, panchayat lands, village commons, raising shelter beds in dry and arid regions and raising trees of different quickly growing species on lands on roadsides, canal banks and railway lines.

Reforestation of degraded forests: situated near and around the villages for the purpose of supplying fuel wood and small timber to villages at fair rates to control encroachments on forests.

Recreation forestry: It is a programme to meet the needs of the urban populations.

Community Forestry:

It is the final expression of the peoples' involvement in the tree plantation, conservation, development and exploitation of forests for the local communities own benefits and for the benefits of the nation. This concept is based on a firm belief in the good sense, responsibility, social conscience and patriotism of even the humblest of the people. It has three dimensions viz.

i)   Restoration or reorganization of the existing forest lands for the total development of both the land and the people therein.

ii) Joint management of forests and its production processes

iii) Development of the socio-economic structure required for the above two.

Community forestry is essentially a value-based, joint management of forestry with satisfaction of the needs, wants and aspiration of both people and the government as its major objective.

Role of Forest Department and NGO's in Conservation:

Conservation of forests is seldom effective, without the support of govt. in the form of large budgets, law enforcement and co-operation between different agencies. So Govt. is having a crucial role to play in the conservation process. Moreover, there should be support of locals as well in the process of conservation. A govt. can provide suitable conditions for participatory forest conservation through the following:

i)   Providing education and working for capacity building

ii) Decentralization of political and administrative powers

iii) Securing the user rights for involved interested groups

The decentralization of power means instead of govt. official's specific groups of interest holders should be given the rights to collect the revenue and decide its proper use. It will strengthen areas where locals are involved in conservation as JFM. If the decentralization of power is proper it will show promising results both in forest protection and local peoples willingness to participate in management and conservation practices.

Conservation requires people's participation and development of the appropriate institutional and regulatory framework for participatory process by the government. For successful forest conservation the optimum formula is joint control and management by government and local people. NGO's also play great role in the process by bringing the two parties on to a common platform to sort out the matters. They bring forth the indifference of govt. to many issues. Presence of capable and environmentally concerned NGO's indicates that changes are taking placed in response to the increasing struggle over the natural resources.

Joint Forest Management:

In India there has been a paradigm shift in terms of forest conservation from an old govt. controlled bureaucratic approach called Traditional Forest Management (in which people were treated as biotic interferences, anthropogenic pressures on forests and it was thought that their involvement will do more harm to the existing forests) to a new/modern approach called Joint Forest Management (which treats people as partners and necessary elements and promotes the symbiotic relationship of forests and the people).

Therefore, the protection, development and regeneration of forests by involving local communities is called a joint forest management (JFM). In India it was recognized for the first time in 1989 by national forest policy, that involvement of people in conservation is necessary for forest management. It holds great promise for developing countries both for resource development and poverty alleviation. Ministry of Environment and Forests (MoEF) Govt. of India (GoI) issued the guidelines for JFM on 1st June 1990 which are as follows:

1) There should be legal back up to the JFM communities under societies registration Act 1960.

2) Participation of women in the joint forest management programme was made necessary and the ratio reserved for them was: 50% of general body members, 33% if executive members and one post of president/ vice president/ secretary.

3) Extension of JFM programme to good forest areas

4) Recognition of self-initiated groups for the conservation of forests under Joint forest management programme.

5) Contribution for resource regeneration: it was stated that 25% of the share of villagers shall be utilized for the objective of resource regeneration.

6) Monitoring and evaluation of the performance of the JFM programme both at divisional/ state level.

Objectives of JFM Programme:

The main objectives of JFM programme are divided into three main parts:

1) Environmental:

   • Protection and maintenance of the fast depleting forest resources.

- To increase thee green cover on the already degraded portion of the forests.

2) Economic:

  - To effectively manage the forests and water resources

  - To offer a means of subsistence and income to the population directly dependent upon the forests

3) Sociopolitical:

  - Empowerment of locals in decision making.

  - Giving the rights to people with specific duties.

4

Human Interaction with Environment

Every living organism influences its environment and in turn gets influenced by it. The magnitude of this influence is not usually high in these species (except humans) because of the fact that due to natural checks, their population cannot rise beyond certain limits and they cannot modify their way of life. However man is an exception. With the increasing scientific knowledge, man is able to modify the environment to suit his immediate needs much more than other organisms. This enables him to modify/ improve the quality of his life. Since the dawn of human civilization man started to interfere with the environment through the following ways:

- He devastated forests by cutting down the trees for wood and other products

- He removed stretches of forests for bringing the land under cultivation, only to find his water resources diminishing and soil resources eroding away

- He killed animals; the gentle ones for food and the fierce ones for his safety

- He polluted the rivers with the chemicals thereby making them unfit for use

- He deteriorated the self-purifying and self-cleaning capacity of the environment by adding heavy loads of pollutants to it.

Our dependence upon the environment is becoming increasing apparent through four interrelated problems given as under:

1) Population Explosion: It will be beyond the capacity of the planet to accommodate so many people. Because the pressure of population mounts the pressure on our limited resources

2) Decline of ecosystems: Humans have significantly altered the critically important ecosystems by disturbing the normal functions of biogeochemical cycles.

3) Global atmosphere changes: Due to anthropogenic enrichment of atmosphere with greenhouses gases the global climate has changed to a large extent.

4) Loss of Biodiversity: Due to the overexploitation of different life forms hundreds of species are completely lost every day leading to the reduction of gene pool.

   Till now the human beings were following the *"utilitarian philosophy"* which states that

all the available resources are for use rather than for show and decoration. But it is high time for all of us to come out of this illogical thinking and play a protective role in our ecological surroundings at least for our own existence.

Environmental Issues and Problems:

The major issues of global concern regarding our environment have been analyzed from time to time by different workers. The World Bank, UNEP and UNDP in the United Nations Conference on Environment and Development (UNCED) have identified four main global issues:

i) Climate change: It includes the greenhouse effect, acid rain formation, smog formation and the atmosphere pollution in general.

ii) Loss of Biodiversity: This is the reduction of the number of species, variety of genes and habitats throughout the world. It also leads to the depletion of gene pool.

iii) International water: The international water resources are threatened by the pollution, droughts and floods. It also includes tmhe siltation of water bodies and erosion of land.

iv) Depletion of Ozone layer: This is the reduction of the thickness of the stratospheric ozone by the anthropogenic activities having consequences on the living organisms of the whole universe.

Besides these global issues the world environment is facing a number of other problems like:

- Population explosion
- Shortage of energy, water and foods
- Unsafe disposal of solid wastes
- Extinction of species
- Deforestation
- Desertification
- Soil erosion and
- Degradation of wetlands

Reasons for the Generation of Environmental Problems:

Basic reasons responsible for the generation of environmental problems and issues are:

i)   Rate of use of the resources is far more than the rate of their renewal

ii) By exceeding the ability of natural systems to assimilate wastes

iii) Causing cumulative degradation of natural and human systems

iv) Irresponsible attitude and character of humans

v)   Continuation of the wrong myths in the society

Solutions to Environmental Problems: Solution to the environmental problems lies in the underlying facts:

i)   Preservation of wilderness through national parks, sanctuaries, biosphere reserves and sacred groves etc.

ii) Sustainable use of resources

iii) Promotion of the concept of sustainable development

iv) Due weightage to the ethical values of environmental protection

v) Promotion of environmental justice: everyone should be given equal rights and opportunities

vi) Mass awareness of people regarding the benefits and values of environment

vii) Proper design of policies and legislations and their proper implementation

viii) Implementation of the decision of international and intergovernmental treaties in the respective countries

ix) Involvement of people in the management of the nature and the resources therein

Values and Beliefs: Values are the basic elements of culture and are of following types:

i) Human values: they show the position of man in nature predicting man in nature rather than nature for man. It clarifies than man is a component of nature and not its master.

ii) Social values: The social values include love, compassion, tolerance, justice and needs to be nurtured by all of us so as to preserve the nature.

iii) Cultural and religious values: These values include the customs, tradition, rituals and the religious concepts which show the sacredness of environmental items.

iv) Ethical values: These are the values which work to change the attitude of humans and make their approach of working as *"Earth Centric"* rather than *"Human Centric"* and are working for the welfare of earth.

v) Global values: Global values show that human civilization is a part of the planet and not the whole planet. They also make it clear that various natural phenomena's are interconnected and interlinked with special bonds of harmony so disturbance of one means disturbance of the whole.

vi) Spiritual values: Spiritual values highlight the principles of self-restraint, self-discipline, contentment, austerity, reduction of wants and freedom from greed i.e. leads to the principle of sustainable use.

Beliefs: Belief means the acceptance of something as true. The basic ideologies of different religious are:

Hinduism: It includes many Vedas like

- Rig-Veda: Believes that universe is made of five basic elements as Earth, Ether, Fire, water and Air.

- Ather-Veda: Believes that nature is the body of God so it is necessary for the followers to work for the betterment of its quality.

- Yajur-Veda: Believes that *"you give me and I give you"* i.e. man should not exploit nature without nurturing it.

Islam: This great religion believes on the beautiful principle of "sustainability". It works for the betterment of every small as well as a big component of the universe. The underlying principles are:

- Simplicity

- Self-restraint

- Non violence

- Contentment

- Discipline

- Conservationism and many more

Buddhism: It lays down emphasis on the conservation by working on the principles of simplicity and ahinsa (non-violence).

Jainism: With the basic principle of simplicity it means minimum destruction of living and non-living resources

Christianity: It is completely baptized in water.

Sikhism: It believes that, *"Air is a vital force, water the progenitor, the vast earth the mother of all, days and night are the nurses fondling all the creatures in their lap"*

Role of Values and Beliefs:

i)   They impose a moral obligation to respect and care for the community and life now and in future

ii) Improve the quality of life by highlighting the importance and sacredness of the components of nature

iii) Conserve the vitality and diversity of earth

iv) Minimize the depletion of non-renewable resources

v)   Change the personal attitude and practices of people

vi) Enable different communities to care for their environment

vii) Formation of a national framework for integrated developed and conservation

viii)Forging of a global alliance for worldwide sustainability and solution of global environmental issues

System Resilience:

Stability is the natural inclination of a system to attain and retain an equilibrium condition. An intrusion into the stability of any system is called disturbance. Every system in nature faces disturbances that can be regular or irregular, short-term or long-term, small scale or large-scale. Small scale disturbances cause a minor damage to the system and get quickly healed up by ecological succession, whereas the large scale disturbances bring about much instability leading to global consequences, which are very difficult to repair and often lead to catastrophes and irreversible disasters. These are called as global disturbances.

System resilience is the ability of a system to recover when it is disturbed. It is also known as the ability of a system to return to its original condition i.e. it is the ability of a system to manage its structure and patterns of behavior in the face of disturbances. Resilient ecosystem is an ecosystem that maintains its normal function and integrity-even through a disturbance e.g. fire appears to be a highly destructive disturbance to a forested landscape. Nevertheless, it releases nutrients that replenish a new crop of plants and in short time burned area is repopulated with trees and is indistinguishable from surrounding area. The process of replenishment of nutrients dispersed surrounding plants and animals, rapid growth of plant cover and succession to a forest can all be thought of as resilience mechanisms. It therefore helps to maintain the sustainability of ecosystems. But it has its limits, which are not to be surpassed.

Biogeochemical Cycles:

A biogeochemical cycle or nutrient cycle is a pathway by which a chemical element or molecule moves through both biotic (biosphere) and abiotic (lithosphere, atmosphere, and hydrosphere) compartments of Earth. All chemical elements occurring in organisms are part of biogeochemical cycles. In addition to being a part of living organisms, these chemical elements also cycle through abiotic factors of ecosystems such as water (hydrosphere), land (lithosphere), and the air (atmosphere). The living factors of the planet can be referred to collectively as the biosphere. All the nutrients such as carbon, nitrogen, oxygen, phosphorus, and sulfur used in ecosystems by living organisms operate on a closed system; therefore, these chemicals are recycled instead of being lost and replenished constantly such as in an open system. The chemicals are sometimes held for long periods of time in one place. This place is called a reservoir, which, for example, includes such things as coal deposits that are storing carbon for a long period of time. When chemicals are held for only short periods of time, they are being held in exchange pools. Examples of exchange pools include plants and animals. Plants and animals utilize carbon to produce carbohydrates, fats, and proteins, which can then be used to build their internal structures or to obtain energy. Plants and animals temporarily use carbon in their systems and then release it back into the air or surrounding medium. Generally, reservoirs are abiotic factors whereas exchange pools are biotic factors. Carbon is held for a relatively short time in plants and animals in comparison to coal deposits. The amount of time that a chemical is held in one place is called its residence. The most well-known and important biogeochemical cycles, for example, include the carbon cycle, the nitrogen cycle, the oxygen cycle, the phosphorus cycle, the sulfur cycle and the water cycle.

Carbon Cycle:

The movement of carbon, in its many forms, between the biosphere, atmosphere, oceans, and geosphere is described by the carbon cycle, illustrated in the diagram. The carbon cycle is one of the biogeochemical cycles. In the cycle there are various sinks, or stores, of carbon and processes by which the various sinks exchange carbon. Plants absorb CO2 from the atmosphere during photosynthesis, also called primary production, and release CO2 back in to the atmosphere during respiration. Another major exchange of CO2 occurs between the oceans and the atmosphere. The dissolved CO2 in the oceans is used by marine biota in photosynthesis. Two other important processes are fossil fuel burning and changing land use. In fossil fuel burning, coal, oil, natural gas, and gasoline are consumed by industry, power plants, and automobiles.

The carbon cycle is based on carbon dioxide (CO2), which can be found in air in the gaseous form and in water in dissolved form. Terrestrial plants use atmospheric carbon dioxide from the atmosphere, to generate oxygen that sustains animal life. Aquatic plants also generate oxygen, but they use carbon dioxide from water.

The process of oxygen generation is called photosynthesis. During photosynthesis, plants and other producers transfer carbon dioxide and water into complex carbohydrates, such as glucose, under the influence of sunlight. Only plants and some bacteria have the ability to conduct this process, because they possess chlorophyll; a pigment molecule in leaves that they can capture solar energy with.

The overall reaction of photosynthesis is:

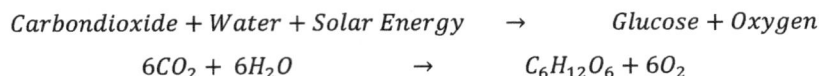

$$Carbondioxide + Water + Solar\ Energy \quad \rightarrow \quad Glucose + Oxygen$$
$$6CO_2 + 6H_2O \quad \rightarrow \quad C_6H_{12}O_6 + 6O_2$$

The oxygen that is produced during photosynthesis will sustain non-producing life forms, such as animals, and most microorganisms. Animals are called consumers, because they use the oxygen that is produced by plants. Carbon dioxide is released back into the atmosphere during respiration of consumers, which breaks down glucose and other complex organic compounds and converts the carbon back to carbon dioxide for reuse by producers.

Carbon that is used by producers, consumers and decomposers cycles fairly rapidly through air, water and biota. But carbon can also be stored as biomass in the roots of trees and other organic matter for many decades. This carbon is released back into the atmosphere by decomposition, as was noted before.

Not all organic matter is immediately decomposed. Under certain conditions dead plant matter accumulates faster than it is decomposed within an ecosystem. The remains are locked away in underground deposits. When layers of sediment compress this matter fossil fuels will be formed, after many centuries. Long-term geological processes may expose the carbon in these fuels to air after a long period of time, but usually the carbon within the fossil fuels is released during combustion processes.

The combustion of fossil fuels has supplied us with energy for as long as we can remember. But the human population of the world has been expanding and so has our demand for energy. That is why fossil fuels are burned very extensively. This is not without consequences, because we are burning fossil fuels much faster than they develop. Because of our actions fossil fuels have become non-renewable recourses.

Although the combustion of fossil fuels mainly adds carbon dioxide to air, some of it is also released during natural processes, such as volcanic eruptions.

In the aquatic ecosystem carbon dioxide can be stored in rocks and sediments. It will take a long time before this carbon dioxide will be released, through weathering of rocks or geologic processes that bring sediment to the surface of water.

Carbon dioxide that is stored in water will be present as either carbonate or bicarbonate ions. These ions are an important part of natural buffers that prevent the water from becoming too acidic or too basic. When the sun warms up the water carbonate and bicarbonate ions will be returned to the atmosphere as carbon dioxide.

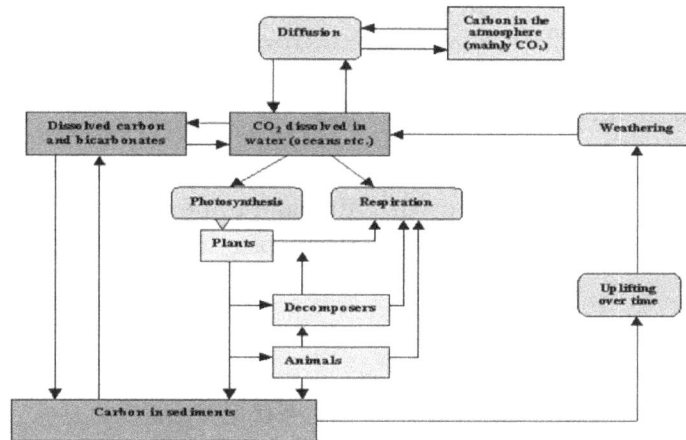

The aquatic carbon cycle

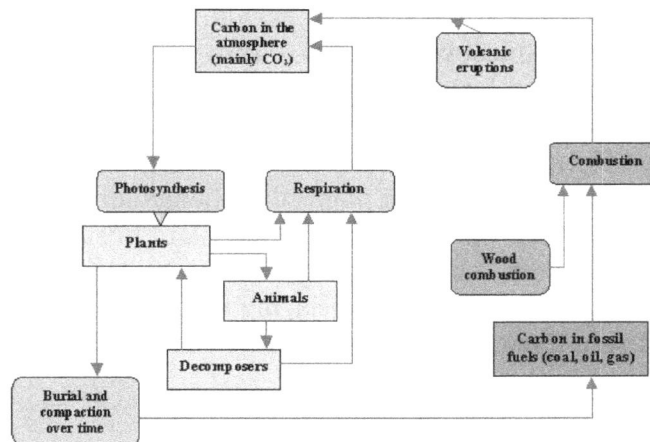

The terrestrial carbon cycle

Human Alteration and Importance

Over the last 150 years, a long-term rise in carbon dioxide has occurred because of humans. The activities that have most contributed to this rise include deforestation and the burning of fossil fuels. Burning coal and oil results in carbon dioxide being released into the air much quicker that it's being removed, resulting in an increase in atmospheric carbon dioxide concentrations.

Due to this increase, the global climate has raised 0.6°C over the past century. If humans continue on with deforestation and consuming fossil fuels, this global warming trend is likely to continue. This could result in a global temperature increase of 1.4 to 5.8 °C over the next 100 years. As temperature rise this will lead to many serious global changes, such as the average sea level rising significantly (0.09 to 0.88 meters) which could result in low-lying cities, such as Portland, New Orleans, Philadelphia, and Washington to severe, frequent floods

Nitrogen cycle: Nitrogen (N) is an essential component of DNA, RNA, and proteins, the building blocks of life. All organisms require nitrogen to live and grow. Although the majority of the air we breathe is N2, most of the nitrogen in the atmosphere is unavailable for use by organisms. Nitrogen is an incredibly versatile element, existing in both inorganic and organic forms as well as in different oxidation states. The movement of nitrogen between the atmosphere, biosphere, and geosphere in different forms is described by the nitrogen cycle one of the major biogeochemical cycles. Similar to the carbon cycle, the nitrogen cycle consists of various storage pools of nitrogen and processes by which the pools exchange nitrogen.

Five main processes cycle nitrogen through the biosphere, atmosphere, and geosphere: nitrogen fixation, nitrogen uptake (organismal growth), nitrogen mineralization (decay), nitrification, and denitrification. Microorganisms, particularly bacteria, play major roles in all of the principal nitrogen transformations.

Nitrogen Fixation

$$N_2 \quad \rightarrow \quad NH^{4+}$$

Nitrogen fixation is the process wherein N2 is converted to ammonium, essential because it is the only way that organisms can attain nitrogen directly from the atmosphere. Certain bacteria, for example those among the genus Rhizobium, are the only organisms that fix nitrogen through metabolic processes. Nitrogen fixing bacteria often form symbiotic relationships with host plants. This symbiosis is well-known to occur in the legume family of plants (e.g. beans, peas, and clover). In this relationship, nitrogen fixing bacteria inhabit legume root nodules and receive carbohydrates and a favorable environment from their host plant in exchange for some of the nitrogen they fix. There are also nitrogen fixing bacteria that exist without plant hosts, known as free-living nitrogen fixers. In aquatic environments, blue-green algae (cyanobacteria) is an important free-living nitrogen fixer. In addition to nitrogen fixing bacteria, high-energy natural events such as lightning, forest fires, and even hot lava flows can cause the fixation of smaller, but significant amounts of nitrogen. Nitrogen uptake

$$NH^{4+} \rightarrow \quad Organic\ Nitrogen$$

The ammonia produced by nitrogen fixing bacteria is usually quickly incorporated into protein and other organic nitrogen compounds, either by a host plant, the bacteria itself, or another soil organism. When organisms nearer the top of the food chain (like us!) eat, we are using nitrogen that has been fixed initially by nitrogen fixing bacteria. Nitrogen mineralization

$$Organic\ Nitrogen \quad \rightarrow \quad NH^{4+}$$

After nitrogen is incorporated into organic matter, it is often converted back into inorganic nitrogen by a process called nitrogen mineralization, otherwise known as decay. When organisms die, decomposers (such as bacteria and fungi) consume the organic matter and lead to the process of decomposition. During this process, a significant amount of the nitrogen contained within the dead organism is converted to ammonium. Once in the form of ammonium, nitrogen is available for use by plants or for further transformation into nitrate (NO_3-) through the process called nitrification.

Nitrification

$$NH^{4+} \rightarrow NO^{3-}$$

Some of the ammonium produced by decomposition is converted to nitrate via a process called nitrification. The bacteria that carry out this reaction gain energy from it. Nitrification requires the presence of oxygen, so nitrification can happen only in oxygen-rich environments like circulating or flowing waters and the very surface layers of soils and sediments. The process of nitrification has some important consequences. Ammonium ions are positively charged and therefore stick (are sorbed) to negatively charged clay particles and soil organic matter. The positive charge prevents ammonium nitrogen from being washed out of the soil (or leached) by rainfall. In contrast, the negatively charged nitrate ion is not held by soil particles and so can be washed down the soil profile, leading to decreased soil fertility and nitrate enrichment of downstream surface and groundwater.

Denitrification

$$NO^{3-} \rightarrow N_2 + N_2O$$

Through denitrification, oxidized forms of nitrogen such as nitrate and nitrite (NO_2-) are converted to dinitrogen (N_2) and, to a lesser extent, nitrous oxide gas. Denitrification is an anaerobic process that is carried out by denitrifying bacteria, which convert nitrate to dinitrogen in the following sequence:

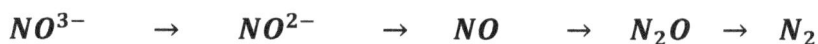

$$NO^{3-} \rightarrow NO^{2-} \rightarrow NO \rightarrow N_2O \rightarrow N_2$$

Denitrification is the only nitrogen transformation that removes nitrogen from ecosystems (essentially irreversibly), and it roughly balances the amount of nitrogen fixed by the nitrogen fixers described above.

Human alteration of the nitrogen cycle and its environmental consequences

Early in the 20th century, a German scientist named Fritz Haber figured out how to short circuit the nitrogen cycle by fixing nitrogen chemically at high temperatures and pressures, creating fertilizers that could be added directly to soil. This technology has spread rapidly over the past century, and, along with the advent of new crop varieties, the use of synthetic nitrogen fertilizers has led to an enormous boom in agricultural productivity. This agricultural productivity has helped us to feed a rapidly growing world population, but the increase in nitrogen fixation has had some negative consequences as well. While the consequences are perhaps not as obvious as an increase in global temperatures or occurrence of a hole in the ozone layer, they are just as serious and potentially harmful for humans and other organisms.

Not all of the nitrogen fertilizer applied to agricultural fields stays to nourish crops. Some is washed off of agricultural fields by rain or irrigation water, where it leaches into surface or ground water and can accumulate. In groundwater that is used as a drinking water source, excess nitrogen can lead to cancer in humans and respiratory distress in infants. The U.S. Environmental Protection Agency has established a standard for nitrogen in drinking water of 10 mg/L nitrate- N. By comparison, nitrate levels in waters that have not been altered by human activity are rarely greater than 1 mg/L. In surface waters, added nitrogen can lead to nutrient over-enrichment, particularly in coastal waters receiving the inflow from polluted rivers. This nutrient over-enrichment, also called eutrophication, has been blamed for increased frequencies of coastal fish-kill events, increased frequencies of harmful algal blooms, and species shifts within coastal ecosystems.

Reactive nitrogen (like NO_3- and NH_4+) present in surface waters and soils, can also enter the atmosphere as the smog-component nitric oxide (NO) and the greenhouse gas nitrous oxide (N_2O). Eventually, this atmospheric nitrogen can be blown into nitrogen-sensitive terrestrial environments, causing long-term changes. For example, nitrogen oxides comprise a significant portion of the acidity in acid rain which has been blamed for forest death and decline in parts of Europe and the Northeast United States. Increases in atmospheric nitrogen deposition have also been blamed for more subtle shifts in dominant species and ecosystem function in some forest and grassland ecosystems. For example, on nitrogen-poor serpentine soils of northern Californian grasslands, plant assemblages have historically been limited to native species that can survive without a lot of nitrogen. There is now some evidence that elevated levels of atmospheric N input from nearby industrial and agricultural development have paved the way for invasion by non-native plants. As noted earlier, NO is also a major factor in the formation of smog, which is known to cause respiratory illnesses like asthma in both children and adults.

Currently, much research is devoted to understanding the effects of nitrogen enrichment in the air, groundwater, and surface water. Scientists are also exploring alternative agricultural practices that will sustain high productivity while decreasing the negative impacts caused by fertilizer use. These studies not only help us quantify how humans have altered the natural world, but increase our understanding of the processes involved in the nitrogen cycle as a whole.

Phosphorus Cycle

Phosphorus is an essential nutrient for plants and animals in the form of ions PO_43- and HPO_42-. It is a part of DNA-molecules, of molecules that store energy (ATP and ADP) and of fats of cell membranes. Phosphorus is also a building block of certain parts of the human and animal body, such as the bones and teeth.

Phosphorus can be found on earth in water, soil and sediments. Unlike the compounds of other matter cycles phosphorus cannot be found in air in the gaseous state. This is because phosphorus is usually liquid at normal temperatures and pressures. It is mainly cycling through water, soil and sediments. In the atmosphere phosphorus can mainly be found as very small dust particles. Phosphorus moves slowly from deposits on land and in sediments, to living organisms, and then much more slowly back into the soil and water sediment. The phosphorus cycle is the slowest one of the matter cycles.

Phosphorus is most commonly found in rock formations and ocean sediments as phosphate salts. Phosphate salts that are released from rocks through weathering usually dissolve in soil, water and will be absorbed by plants. Because the quantities of phosphorus in soil are generally small, it is often the limiting factor for plant growth. That is why humans often apply phosphate fertilizers on farmland. Phosphates are also limiting factors for plant-growth in marine ecosystems, because they are not very water-soluble. Animals absorb phosphates by eating plants or plant-eating animals.

Phosphorus cycles through plants and animals much faster than it does through rocks and sediments. When animals and plants die, phosphates will return to the soils or oceans again during decay. After that, phosphorus will end up in sediments or rock formations again, remaining there for millions of years. Eventually, phosphorus is released again through weathering and the cycle starts over.

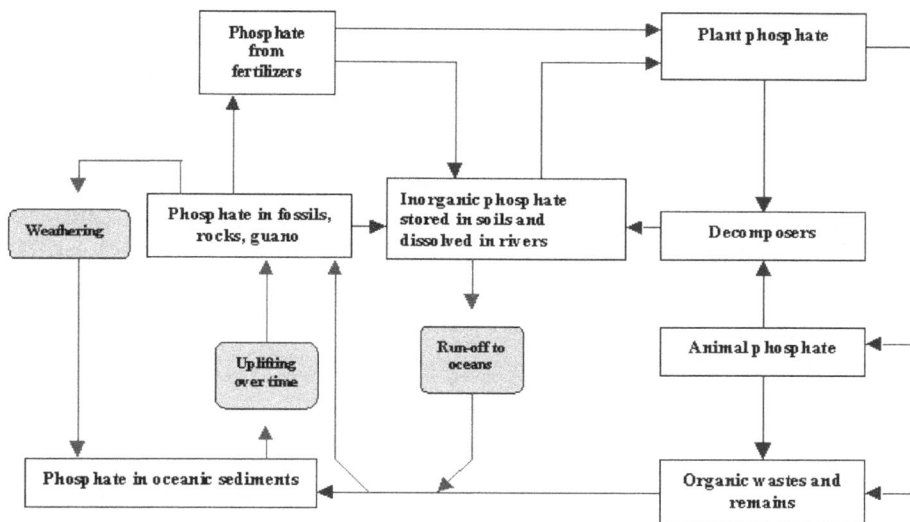

A schematic representation of the phosphorus cycle

Human Inputs to the Phosphorus Cycle:

Human influences on the phosphate cycle come mainly from the introduction and use of commercial synthetic fertilizers. The phosphate is obtained through mining of certain deposits of calcium phosphate called apatite. Huge quantities of sulfuric acid are used in the conversion of the phosphate rock into a fertilizer product called "super phosphate". Plants may not be able to utilize all of the phosphate fertilizer applied, as a consequence, much of it is lost form the land through the water run-off. The phosphate in the water is eventually precipitated as sediments at the bottom of the body of water. In certain lakes and ponds this may be redissolved and recycled as a problem nutrient.

Animal wastes or manure may also be applied to the land as fertilizer. If misapplied on frozen ground during the winter, much of it may lose as run-off during the spring thaw. In certain areas very large feed lots of animals, may result in excessive run-off of phosphate and nitrate into streams. Other human sources of phosphate are in the out flows from municipal sewage treatment plants. Without an expensive tertiary treatment, the phosphate in sewage is not removed during various treatment operations. Again an extra amount of phosphate enters the water.

Sulphur Cycle:

Sulphur is one of the components that make up proteins and vitamins. Proteins consist of amino acids that contain sulphur atoms. Sulphur is important for the functioning of proteins and enzymes in plants, and in animals that depend upon plants for sulphur. Plants absorb sulphur when it is dissolved in water. Animals consume these plants, so that they take up enough sulphur to maintain their health.

Most of the earth's sulphur is tied up in rocks and salts or buried deep in the ocean in oceanic sediments. Sulphur can also be found in the atmosphere. It enters the atmosphere through both natural and human sources. Natural recourses can be for instance volcanic eruptions, bacterial processes, evaporation from water, or decaying organisms. When sulphur enters the atmosphere through human activity, this is mainly a consequence of industrial processes where sulphur dioxide (SO2) and hydrogen sulphide (H2S) gases are emitted on a wide scale. When sulphur dioxide enters the atmosphere it will react with oxygen to produce sulphur trioxide gas (SO3), or with other chemicals in the atmosphere, to produce sulphur salts. Sulphur dioxide may also react with water to produce sulphuric acid (H2SO4). Sulphuric acid may also be produced from demethylsulphide, which is emitted to the atmosphere by plankton species. All these particles will settle back onto earth, or react with rain and fall back onto earth as acid deposition. The particles will then be absorbed by plants again and released back into the atmosphere, so that the sulphur cycle will start over again.

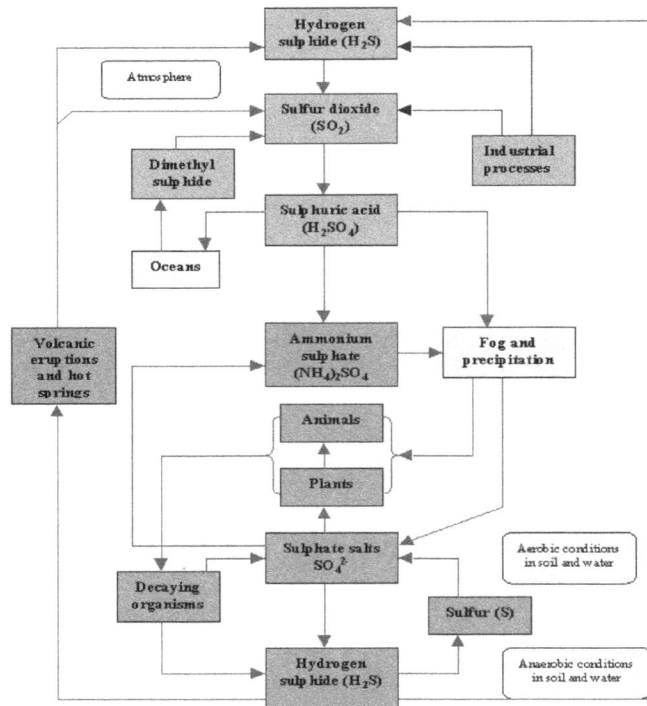

A schematic representation of the sulphur cycle

Human Activities

Sulphur is also one of the main biogeochemical cycles that is significantly influenced by human activities. According to some estimates, emissions of sulphur to the atmosphere from human activities are at least equal or probably larger in magnitude than those from natural processes. Burning of fossil fuels and metal processing are the main culprits. Sulphuric acid particles contribute to the polluting smog that engulfs some industrial centers and cities where many sulphur containing fuels are burned. Sulphur dioxide is also involved in the phenomenon of acid rain. Such particles floating in air (known as sulfate aerosols) can cause respiratory diseases or cool the climate by reflecting some extra sunlight to space.

The levels of sulphur dioxide can be reduced by controlling pollution in coal-based power plants and in industrial processing units. Use of fuels like natural gas which have lower sulphur content also helps.

Fossil Fuels and their Impact on the Environment:

The technical definition of fossil fuels is "incompletely oxidized and decayed animal and vegetable materials, specifically coal, peat, lignite, petroleum and natural gas" or "material that can be burned or otherwise consumed to produce heat". In our modernized western world, fossil fuels provide vast luxurious importance. We retrieve these fossil fuels from the ground and under the sea and convert them into electricity. Approximately 90% of the world's electricity demand is generated from the use of fossil fuels.

There is a growing concern regarding the collaboration between fossil fuels and environmental pollution. Debates regarding this contamination have become commonplace in today's effort to sustain the earth's health. Fossil fuels are not considered a renewable energy source and aside from the environmental impact, the cost of retrieving and converting them is beginning to demand notice. Seemingly this issue has many different angles that need to be addressed in order to ensure future generations a sustainable living.

Combustion of these fossil fuels is considered to be the largest contributing factor to the release of greenhouse gases into the atmosphere. In fact it is believed that energy providers are the largest source of atmospheric pollution today. There are many types of harmful outcomes which result from the process of converting fossil fuels to energy. Some of these include air pollution, water pollution, accumulation of solid waste, not to mention the land degradation and human illness.

Evidence of the ill effects of fossil fuels is endless, and can take on many forms. Some forms are not easily seen by the human eye, although the disastrous results such as the loss of aquatic life can be seen somewhat after the fact. Carbon dioxide is considered the most prominent contributor to the global warming issue. The impact of global warming on the environment is extensive and affects many areas. In the Antarctica, warmer temperatures may result in more rapid ice melting which increases sea level and compromises the composition of surrounding waters. Rising sea levels alone can impede processes ranging from settlement, agriculture and fishing both commercially and recreationally.

Air pollution is another problem arising from the use of fossil fuels, and can result in the formation of smog causing human problems and affecting the sustainability of crops. Smog seeps through the protective layer on the leaves and destroys essential cell membranes. This results in smaller yields and weaker crops, as the plants are forced to focus on internal repair and do not thrive. Many toxic substances are released during the conversion or retrieval process including "Vanadium" and "Mercury". According to the "New Book Of Popular Science", "it is suspected that significant quantities of Vanadium in the atmosphere results from residual fuel oil combustion".

When coal is burned, it releases nitrous oxide. Unfortunately this is kept in the atmosphere for very long time. The harmful impact of this chemical could take up to a couple of hundred years to make itself known. It is very difficult to prevent or to diminish an impact when you are not even aware of what it may be. The only solution in this case is to reduce the formation of nitrous oxide. Nearly 50% of the nitrogen oxide and 70% of sulfur dioxide in the atmosphere are direct result of emissions released when coal is burned.

Converting fossil fuels may also result in the accumulation of solid waste. This type of accumulation has a devastating impact on the environment. Waste requires adequate land space for containment and/or treatment, as well as financial support and monitoring for wastes not easily disposed of. This type of waste also increases the risk of toxic runoff which can poison surface and groundwater sources for many miles. Toxic runoff also endangers surrounding vegetation, wildlife, and marine life.

Delivery of fossil fuels can result in oil spills, and many of us are familiar with the impacts of this type of disaster. Seepage from foundations like that of oil rigs and pipelines can also result in similar destruction for habitat and wildlife. According to the Department Of The Interior, vast damage to waterways can be attributed to the extraction of coal. Coal extraction may very well be the leading the source of water pollution today.

Use of unleaded gas has helped to reduce the release of lead into the environment. Although in third world countries, the safer unleaded gas has not been fully utilized and is still a major concern. Unfortunately for developing countries, the economy and technology available to them is quite behind what we are used to. With this in mind many environmental issues are treated at an international level, which allows for more efficient handling.

We have become a very energy greedy generation and our demands for electricity are very high. As far as reducing these harmful effects, we must first reduce our demand. Science may be able to find alternative, healthier sources, although not ones that meet the required supply. These types of horrendous impacts are felt globally and should not be considered one countries problem. Sometimes social limitations and/or economic stability can make the process of change very difficult. One thing is for sure, that by being more energy efficient and conservative, we will be helping to alleviate the toll on environmental and human health.

Water Cycle

Water cycle, also known as the hydrologic cycle describes the continuous movement of water on, above and below the surface of the Earth. Water can change states among liquid, vapor and ice at various places in the water cycle. Although the balance of water on Earth remains fairly constant over time, individual water molecules can come and go, in and out of the atmosphere. The planetary water supply is dominated by the oceans. Approximately 97 % of all the water on the Earth is in the oceans. The other 3 % is held as freshwater in glaciers and icecaps, groundwater, lakes, soil, the atmosphere, and within life. Water is continually cycled between its various reservoirs. This cycling occurs through the processes of evaporation, condensation, precipitation, deposition, run-off, infiltration, sublimation, transpiration, melting, and groundwater flow. The water moves from one reservoir to another, such as from river to ocean, or from the ocean to the atmosphere, by the physical processes of evaporation, condensation, precipitation, infiltration, runoff, and subsurface flow. In doing so, the water goes through different phases viz. are solid, liquid, and gas.

It is the circulation of water within the earth's hydrosphere, involving changes in the physical state of water between solid, liquid, and gas phases and refers to the continuous exchange of water between atmosphere, land, surface and subsurface waters, and organisms. In addition to storage in various compartments like ocean the multiple cycles that make up the earth's water cycle involve five main physical actions: *evaporation, precipitation, infiltration, runoff, and subsurface flow.*

Evaporation: Evaporation is movement of free water to the atmosphere as a gas. It requires large amounts of energy. It occurs when radiant energy from the sun heats water, causing the water molecules to become so active that some of them rise into the atmosphere as vapour. This transfer entails a change in the physical nature of water from liquid to gaseous phases. Along with evaporation can be counted transpiration from plants. Thus, this transfer is sometimes referred to as evapotranspiration. About 90% of atmospheric water comes from evaporation, while the remaining 10% is from transpiration. Transpiration occurs when plants take in water through the roots and release it through the leaves, a process that can clean water by removing contaminants and pollution. Evapotranspiration is water *evaporating* from the ground and *transpiration* by plants. Evapotranspiration is also the way water vapour re-enters the atmosphere.

Precipitation: Rising warm air carries water vapor high into the sky where it cools, forming water droplets around tiny bits of dust in the air. Some vapor freezes into tiny ice crystals which attract cooled water drops. The drops freeze to the ice crystals, forming larger crystals called as snowflakes. When the snowflakes become heavy, they fall and when they meet warmer air on the way down, they melt into raindrops. They bang together and grow in size until they are heavy enough to fall. Sometimes there is a layer of air in the clouds that is above freezing. Then closer to the ground the air temperature is once again below freezing. Snowflakes partially melt in the layer of warmer air, but then freeze again in the cold air near the ground. This kind of precipitation is called sleet. There is another kind of precipitation that comes from thunderstorms called hail.

Infiltration: Under some circumstances precipitation actually evaporates before it reaches the surface. More often, though, precipitation reaches the Earth's surface, adding to the surface water in streams and lakes, or infiltrating. A portion of the precipitation that reaches the Earth's surface seeps into the ground through the process called infiltration. Infiltration into the ground is the transition from surface water to groundwater. The infiltration rate depends upon soil or rock permeability as well as other factors. Infiltrated water reaches another compartment known as groundwater (i.e., an aquifer). Groundwater tends to move slowly, so the water returns as surface water after storage within an aquifer for a period of time that can amount to thousands of years in some cases.

Runoff: The amount of water that infiltrates the soil varies with the degree of land slope, the amount and type of vegetation, soil type and rock type, and whether the soil is already saturated by water. The more the openings in the surface (cracks, pores, joints), the more is the infiltration.

Water that doesn't infiltrate the soil flows on the surface as runoff. Precipitation that reaches the surface of the Earth but does not infiltrate the soil is called runoff. Runoff can also come from melted snow and ice. It also includes the variety of ways by which land surface water moves down slope to the oceans. Water flowing in streams and rivers may be delayed for a time in lakes. Not all precipitated water returns to the sea as runoff; much of it evaporates before reaching the ocean or reaching an aquifer.

Subsurface Flow: Surface flow incorporates movement of water within the earth, either within the recharge zone or aquifers. After infiltrating, subsurface water may return to the surface or eventually seep into the ocean.

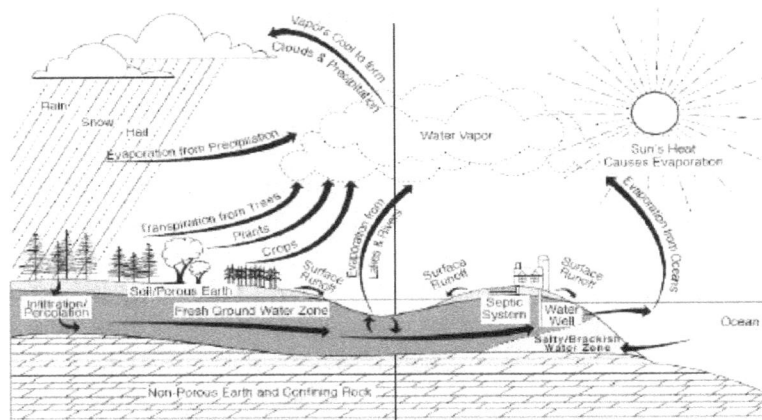

Hydrological Cycle

5

Atmospheric Pollution

Recently for the first time in his cultural history, man has force one of the most horrible the problem of pollution of his environment which sometimes in part was pure, virgin, undisturbed, uncontaminated and basically quite hospitable for him. Pollution is an undesirable change in the physical, chemical or biological characteristics of our environment (air, water and land) that may harmfully affect human life or that of other species, living conditions and cultural assets. In other words, pollution is the unfavorable alteration of the environment, largely as a result of human activities. It also means the presence of undesirable substances in any segment of environment primarily due to human activities discharging waste products or harmful secondary products which are harmful to man and other living organisms.

Air Pollution:

The presence of one or more contaminants such as dust, smoke, smog or vapors in the outdoor atmosphere in quantities or characteristics and of duration so as to be injurious to human plant and animal life and property or which unreasonably interferes with the comfortable enjoyment of life and property is known as air pollution *(US Public Health services)*

The atmosphere is a dynamic system, which steadily absorbs various pollutants from natural as well as man-made sources, thus acting as a natural sink. Gases such as CO_2 CO, H_2S, oxides of sulpher (SOX) and oxides of nitrogen (NOX) as well as particulate matter like dust and sand are continuously released into the atmosphere through natural activities as forest fires, volcanic eruptions, decay of vegetation, wind and dust storms. Man-made pollutants e.g. CO_2, NOX, CO, hydrocarbons, CFC's, particulates etc. are also released into the atmosphere. These have surpassed the pollutants contributed by nature thousand fold. The magnitude of the problem of atmospheric pollution has increased dramatically due to population explosion, industrialization, urbanization, automobiles and other human activities. If the rate of entry of these pollutants is faster than the rate of absorption of these pollutants by the atmosphere (Natural Sink) then they gradually accumulate in air. Such disturbance in the dynamic equilibrium of the atmosphere by the air pollutants released by human activities resulting in considerable accumulation in the atmosphere may affect the life on earth and its environment.

Classification of Air Pollutants: The air pollutants can be classified in different types on the following basis.

    i)   According to Origin:

        a)   Primary Pollutants: Pollutants which are directly emitted into the atmosphere and are found as such in it e.g. CO and SO_2.

        b)   Secondary Pollutants: Pollutants which are derived from primary pollutants due to chemical or photochemical reactions in atmosphere e.g. PAN and O_3.

ii) According to Chemical Composition:

   a) Organic Pollutants: These are the compounds containing carbon and hydrogen and may also contain elements like oxygen, nitrogen, phosphorus and sulphur. Hydrocarbons, aldehydes, ketones and other organic compounds of concern in the air pollution field are carboxylic acids, alcohols, ethers, esters, amines and organic sulphur compounds.

   b) Inorganic Pollutants: These include carbon compounds like carbon monoxide (CO), nitrogen compounds like oxides of nitrogen (NOX), ammonia (NH3), sulphur compounds like hydrogen sulphide (H2S), sulpher dioxide (SO2), sulpher trioxide (SO3) and sulphuric acid (H2SO4).

iii) According to states of matter:

   a) Gaseous Pollutants: Pollutants that get mixed with air and do not settle down are called as gaseous pollutants. CO, CO2, SOX, and NOX are a few examples of gaseous pollutants.

   b) Particulate Pollutants: pollutants comprising of finely divided solids and liquids often present in colloidal state as aerosols as smoke, fumes, dust, fog, sprays and fly ash are called as particulate pollutants.

Sources of Air pollution: the sources of air pollution have been divided into two general types

1) Natural Sources: these sources include

   o Volcanic eruptions

   o Forest fires

   o Natural decay of vegetation

   o Wind storm

   o Dust storm

   o Marsh gases

   o Defilation of sand and dust

   o Pollen grains of flowers

2) Man-made/Anthropogenic Sources: these sources include

   o Deforestation

   o Burning of fossil fuels

   o Emissions of vehicles

   o Rapid industrialization

o   Agricultural activities

o   Wars

Control of Air Pollution:

air pollution can be controlled by applying the following two approaches.

1. Controlling or confining the pollutants at sources: Pollutants can be controlled at source either by modifying the process in such a way that pollutants are not formed at all beyond the permissible limits or by reducing the pollutant concentration to tolerate limits before they are released into the atmosphere by the use of suitable equipments to destroy, alter or trap the pollutants formed.

2. Dilution of the pollutants: It is the reduction of the concentration of pollutants in the atmosphere to permissible level. It can be achieved by using tall stacks, controlling the process with due regard to the local meteorological conditions and proper community planning to prevent accumulation of dangerous ground level concentrations within the designated areas.

General methods of Air pollution control: Following are the general methods of air pollution control.

i)   Zoning: Zoning of industries is done on the basis of type and function of industry. If zoning is done properly, it results in considerable improvement of health of the community. As a whole it prevent the invasion of undesirable pollutants of industries in an around residential areas. So, toxic, hazardous, harmful gases and odors are prevented from entering or attacking the human life in residential areas. The industry causing nuisance and producing undesirable gases and odors and other toxic products may be located away from towns in spacious lands.

ii) Control at Source: It can be done by making use of following measures.

   a) Substitution of raw material

   b) Modification of process

   c) Alteration in the equipments.

iii) Installation of controlling devices and equipments: There are two categories of devices which are often used for air pollution control including those used for reducing particulate matter and gaseous pollutants.

   a) Control of particulate pollutants:

i)   Gravitational settling chambers: These are the devices in which the velocity of horizontal carrier gas is reduced adequately so that particles settle down by gravitational forces. Particles with diameter ranging between 40µm and 100µm are readily collected by this technique. The usual velocity through setting chambers is between 0.5 and 2.5m/s, although for best results the gas flow should be uniformly maintained at less than 0.3m/s. The efficiency of these chambers is very poor on fine particles and decreases as the load increases.

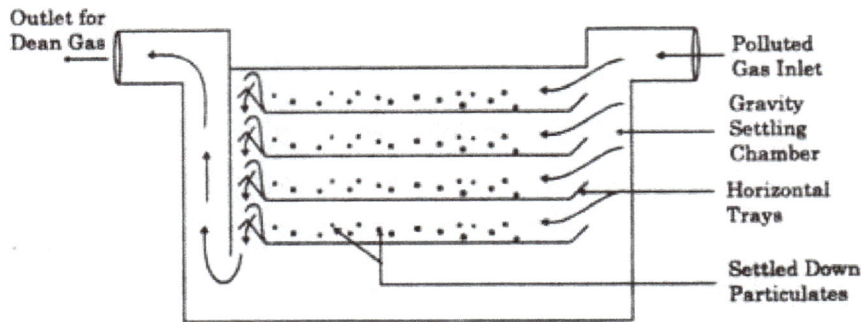

Fig.1. Gravitational Settling Chamber

ii) Cyclonic collectors: A cyclonic collector is a device consisting of a cylindrical shell, conical base, dust hopper and an inlet where the dirty air enters tangentially. In this device the velocity of the incoming gas is transformed into a vortex from which the centrifugal force drives the suspended particles to the walls of the collector. Thus the sudden change in the direction of gas flow causes the particles to separate out due to their greater momentum. Cyclonic collectors are used to remove particulates from rock product industries, iron industries, steel plants, mining and metallurgical industries. They are best for collection of particles of size 15-50 μm. The efficiency of the cyclonic collectors depends on the magnitude of the centrifugal force exerted on the particles. The greater the centrifugal force, the greater is the separating efficiency. The magnitude of the centrifugal force generated also depends on the particle mass, gas velocity within the cyclone and cyclonic diameter.

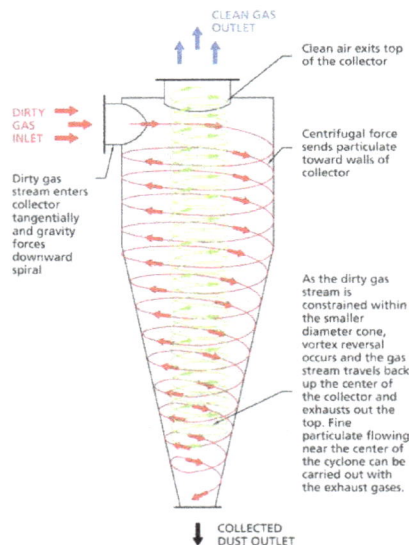

Fig.2. Cyclonic Collector

iii) Dynamic precipitators: These devices also work on the principle of centrifugal force. The centrifugal force generated by the rotating blade probes the particles in air stream from where they are drawn off in a concentrated stream. Particles of size between 5-20 μm are readily collected by these devices. However, the devices cannot handle wet fibrous materials which get accumulate on their blades.

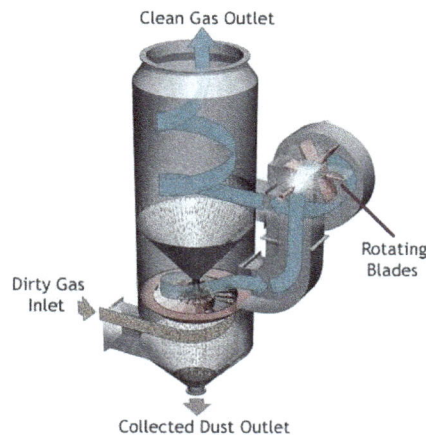

Fig.3. Dynamic precipitator

iv) Wet Collection devices: Devices using mixed phase of gases and liquids are known as wet washers and scrubbers. These are the collection devices in which the particles are washed out of the gas flow by a water spray. The most common wet collection devices used are:

a) Cyclonic Scrubber: In this, the aerosol is introduced in a centrifugal manner. Water is sprayed at the internee of the gas and plates are provided to remove the moisture from the gas after the removal of dust. This is followed by control equipment like gravity setting chamber or cyclonic precipitator. It can remove dust particles of 5 µm size with an efficiency of 90%.

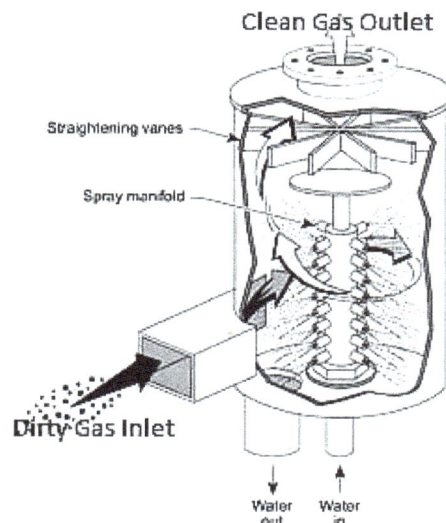

Fig.4. Cyclonic Scrubber

b) Venturi Scrubber: Small particles of size 0.5 to 5 µm associated with smoke and fumes are effectively removed by highly efficient venture scrubbers. The polluted gas (at a velocity of 60 -180m/s) passes through a duct that has a venturi-shaped throat section. A coarse water spray is injected into the throat, where it is atomized by the high gas velocity. The particulates in the gas stream collide with the liquid droplets and get entrained in the droplets and fall down for removal. Venture scrubbers can also remove soluble gaseous contaminants. The efficiency of these instruments is more than 90%.

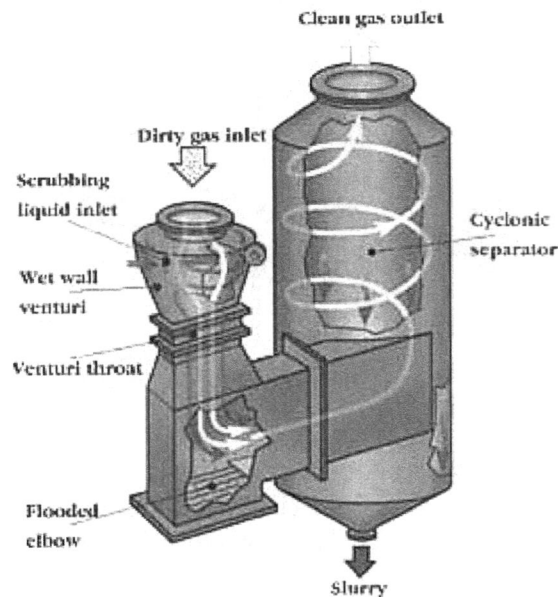

Fig.5. Venturi Scrubber

v) **Electrostatic precipitator (ESP):** When a gas stream contains aerosols e.g. dust, fumes or mist of diameter as small as 0.0001 cm, it can be passed through a more versatile and efficient type of device called electrostatic precipitator. It works on the principle that when the particulates move through a region of high electric potential (30,000 -1,00,000 V) they become charged and are attracted towards oppositely charged areas where they are collected and removed. The ESP consists of a series of plates which are charged to high voltages alternatively (+) positive and (-) negative. Particles approaching a given plate acquire its charge and are then attracted to the surface of next plate from where they fall into the hopper below. Thus particles pick up charge as they pass between plates and are precipitated on plates of opposite charge. Four basic steps are involved in the operation of an Electrostatic precipitator.

- Electrically charging of particles by ionization.

- Transporting the charged particles to the collecting surface.

- Neutralizing the electrically charged particles precipitated on the collecting surface.

- Removing the precipitated particles from the collecting surface by washing.

Electrostatic precipitators prove to be the valuable devices when:

- Very large volumes of gases are to be handled.

- Valuable dry material is to be removed.

- The gas temperature is very high.

They are highly efficient with their efficiency approaching 99.9% and are widely used in power plants, cement industries, paper and pulp industries, iron and steel industries etc.

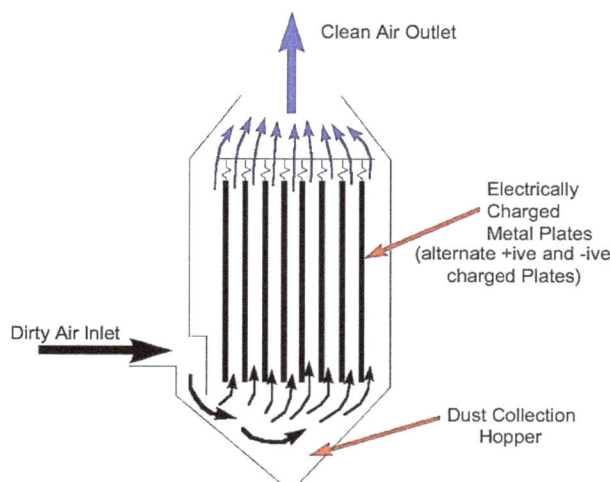

Fig.6. Electrostatic precipitator

vi) Filter Systems: In the fabric filter systems the dust laden gases are forced to pass through a porous medium as woven fabric. The particulates are trapped and collected in the filters and the gases devoid of the particulates are discharged out. Fibrous or deep-bed filters and cloth bag-filters, nylon, Dacron, asbestos, silicon coated glass cloth etc. are also used. Cloth and nylon filters are used when the temperature is up to 80-90°C whereas asbestos and silicon coated glass cloth is used at a temperature of 250-350°C. Wool filters are good for acidic gases whereas cloth, nylon and asbestos are good for alkaline gases. These provide a very efficient method for the removal of particulates even in a range less than 0.5 μm diameter with an efficiency of 99%. The fabric must be cleaned frequently by blowing compressed gas in the reverse direction otherwise no gas will be able to pass through it.

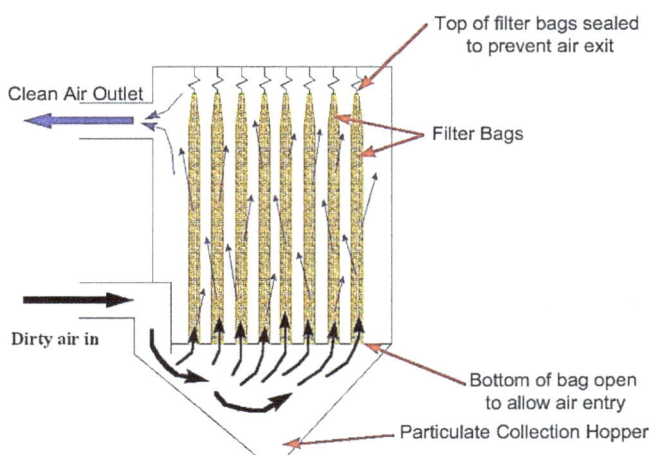

Fig. 7. Bag House Filter

b) Control of gaseous pollutants: The main gases of concern in air pollution control are the oxides of sulphur, carbon and nitrogen, organic and inorganic acidic gases and hydrocarbons. Major treatment processes currently available for control of these and other gaseous emissions include.

i)   Absorption

ii) Adsorption

iii) Combustion

iv) Condensation

v)   Masking and counter acting

i)   Absorption: Absorption or scrubbing, involves bringing contaminated effluent gas (absorbate) in contact with liquid absorbent so that one or more constituents of the effluent gas are removed, treated or modified by the liquid absorbent. The absorbent may utilize either chemical or physical change in removing pollutants. The efficiency of the process depends upon.

   • Chemical reactivity of the gaseous pollutants in the liquid phase.

   • Extent of surface contact between the liquid and the gas.

   • Contact time and

   • Concentration of absorbing medium.

The equipments used include plate towers, spray towers, packed towers, bubble cap plate towers and liquid fit scrubber towers. The gas absorption technique is widely used for removing pollutants like NOX, H2S, SO2, SO3 and fluorides from gaseous effluent. The various absorbing techniques use the following liquids.

Pollutant	Absorbent
NO_x	Nitric acid, water
HF	Water, sodium hydroxide
H_2S	Ethanol amine, NaOH + Phenol (3:2)
SO_2	Water, alkaline water

ii) Adsorption: It involves the passage of a stream of effluent gas through a porous solid material (adsorbent) contained in an adsorption bed. The technique of gas absorption is based on the retention of gases on solid adsorbents. The adsorption may be physical or chemical. Physical adsorption is dependent upon the temperature and pressure conditions in the system. It is promoted by increase in pressure and decrease in temperature. It depends upon Vander Waals force (intermolecular attractive force). Chemical adsorption depends upon the reactivity of gases and their bond forming capacity with the surface of adsorbent. The adsorption materials commonly used are solids such as silica gel, activated alumina and activated charcoal. When the waste gas stream contains higher concentration of gases like SOX and NOX, they can be recovered economically and used for the manufacture of H2SO4 and HNO3 respectively. The pollutants from power plants can be removed by injecting pulverized lime stone into the boiler furnace. The CaO formed reacts with SOX to form calcium sulphate and calcium sulphide, thus SOX emission in the atmosphere is prevented.

Pollutant	Absorbent
NOX	Silica gel, commercial ziolitic
HF	Porous pellets of Na, Lime stone
H2S	Iron Oxide
SO2	Pulverized lime stone or dolomite
Organic vapours	Activated carbon
Petroleum	beuxite

iii) **Masking and Counter Acting:** This is a control method chiefly for odors, by which they are suppressed in masks by the addition of some pleasant odor producing substance. The pleasant odor producing substances should not be toxic, corrosive or allergic e.g. addition of vanilla flavors or other flavors to the primary clarifiers of a sewage treatment plant masks the odors of H_2S (hydrogen sulphide) and CH_4 (methane).

iv) **Combustion:** Though it is a major source of air pollution, combustion or incineration is also the basis for an important air pollution control process in which the objective is to convert the air contaminants (usually hydrocarbons or CO) to carbon dioxide (CO_2) and water. For efficient combustion to occur, it is necessary to have the proper combination of four basic elements.

i) Oxygen supply

ii) Temperature

iii) Turbulence

iv) Time

Depending upon the contaminant being oxidized, direct flame combustion, thermal combustion or catalytic combustion methods can be used to control air pollution.

Green House Effect and Global Warming

Greenhouse means a building made mainly of glass, with heat and humidity regulated for growing plants. Solar radiation passes through the glass and the inside environment is heated up. The heat inside the house is contained within as the heated air does not leave the house. Consequently, the glass house will be warmer than the outside atmosphere. In real atmosphere, the heat energy is contained by certain gases and hence, called *Green house effect*. During daytime, solar energy passes through the atmosphere and reaches the surface of the earth. During its passage, parts of the incoming solar radiation are scattered, reflected, or absorbed. The remaining radiation reaches the earth's surface and thereby the earth's surface is heated up. The energy received by the earth's surface throughout the daytime heats up the earth's surface continuously. As the earth's surface is heated up, it will become hot and hold the thermal energy. The earth at this point, starts emitting energy in the form of long wave radiation called *outgoing radiation*.

Part of this outgoing radiation is absorbed by the atmosphere and retained as heat energy. The remaining energy escapes into the outer spaces. Carbon dioxide, water vapor, and few other gases in the atmosphere are capable of absorbing this outgoing radiation. The concentration of these gases (called, green house gases) determines the amount of radiations absorbed. More the concentration of these gases, more the amount of long wave radiations absorbed. It is a natural phenomenon occurring in the atmosphere. It helps preventing the earth from drastic cooling of its surface and its atmosphere. If all the heat energy from the earth's surface is lost to the space, then earth will become too cold to support the life. These gases act like a blanket and provide the earth with "blanketing effect". This process is called Green house effect. In nutshell, atmosphere, like glass absorbs some of the long wave radiation emitted by earth and radiates the energy back to the earth. In this way temperature of the earth is maintained.

Global Warming

The four major greenhouse gases, which cause adverse effects, are carbon dioxide (CO_2), methane (CH_4), nitrous oxide (N_2O) and chlorofluorocarbons (CFCs). Among these CO2 is the most common and important greenhouse gas. Here it should be noted that ozone and SO_2 also act as serious pollutants in causing global warming. The other greenhouse gases such as methane and chlorofluorocarbons contribute about 18% and 14% respectively to the global warming. Since the beginning of the industrial revolution, three human activities have contributed to significant rise of concentration of green house gases. The activities are:

1. Use of fossil fuels: it releases huge amounts of carbon dioxide into the atmosphere.

2. Deforestation and burning of forests and grasslands to convert into cropland: forests and grasslands are cleared and/or burned for converting them into cropland. Burning of biomass produces large quantities of CO_2. Clearance of forests also deprives available vegetation for absorption of CO_2 through photosynthesis.

3. Cultivation of rice in paddies and use of inorganic fertilizers: cultivation of rice in paddies generates methane and use of fertilizers release N_2O into the atmosphere. Since, 1860, the concentration of green house gases, CO_2, CH_4 and N_2O have risen sharply especially since, 1950. Burning of coal for power generation and for industrial purposes and burning of petroleum products by millions of vehicles are the two major contributors of CO_2 emissions.

As the green house gases increase dramatically, by human activities, the green house effect is enhanced. That is more amount of outgoing long wave radiation is retained than required amount. It results in greater warming of the atmosphere than normal. It is called "enhanced green house effect".

Due to this enhancement, the earth's atmosphere is warming up gradually more and more. This phenomenon is called global warming. Since 1960 total atmospheric carbon dioxide has increased from about 320 to over 350 ppm. Over the same period the average global temperature has increased very slightly, 0.6 ± 0.2 °C. Nine warmest years have occurred since 1990. The hottest year was 1998, followed in order by 2002 and 2001. There is an apparent correlation between increases in fossil fuel use, atmospheric concentration of CO_2 and global temperature between 1970 and 2002.

At earth's poles and in Greenland, the temperature rise has been noticed, so also some melting of land-based ice caps and floating ice. Some glaciers on the top of mountains in Alps, Andes, Himalayas, northern Cascades of Washington, and Mount Kilimanjaro in Africa have begun shrinking due to melting of ice.

In addition, some warm-climate fish and other species have migrated northward. The season, spring arrives earlier and autumn arrives later than normal in many parts of the globe. All the above suggest that the global warming is real and happening. If the warming continues, the ice from polar region and from mountain glacier will melt (not all the ice, but some) which will cause coastal flooding. Coastal flooding will result in rise of the sea level and consequently inundation of coastal areas (villages, towns, cities). It is estimated that sea level will rise by 1.5 m by 2050.

Ozone Layer Depletion

Today, due to human activities, this ozone layer is becoming thin. This thinning is called, ozone depletion. At the zones, where thinning is too severe, they are termed as "Ozone holes". The ozone hole is defined as the area having less than 220 Dobson units (DU) of ozone in the overhead column (i.e., between the ground and space).

Mechanism of ozone depletion is not well understood. Chlorofluorocarbons (CFC) used in refrigerators, air conditioners, propellants etc. and oxides of nitrogen emitted by aircrafts flying near stratosphere are found to be the causes for ozone layer depletion. Ozone layer is depleted by free radical catalysts – nitric oxide (NO), hydroxyl (OH), atomic chlorine (Cl), and atomic bromine (Br). Halogens have the ability to catalyze ozone breakdown.

A catalyst is a compound which can alter the rate of a reaction without permanently being altered by that reaction and so can react over and over again. Chlorine atom or any other halogen atom released from CFC or BFC by striking ultraviolet radiation is now available for catalyzing ozone breakdown. Although these species occur naturally, large amounts are released by human activities through the use of CFCs and bromofluorocarbons.

Ozone Layer Depletion by CFC's

CFCs and BFCs are stable compounds and live long in the atmosphere. The Cl and Br radicals are liberated from these compounds by the action of ultraviolet radiation. These radicals initiate and catalyze breaking the ozone molecules. One single radical is capable of breaking down over 1,00,000 ozone molecules. Ozone concentration is decreasing at the rate of 4% per decade over northern hemisphere. CFCs have long life time – 50 to 100 years. As they remain for such a long duration, they deplete ozone layer continuously. Moreover, this depletion rate keeps increasing as more and more CFCs are released.

After realizing the seriousness of this problem, countries have come forward to ban completely or phase out use and manufacture of CFCs. Sweden was the first nation to ban CFC-containing aerosol sprays. The ban was brought to force on January 23, 1978 in Sweden. Few other countries followed Sweden later that year. Ozone hole was discovered in the year 1985. After this, countries came forward for an international treaty, the Montreal Protocol for complete phase-out of CFCs by 1996. This effort has yielded positive result. Scientists announced on August 2, 2003 that depletion of ozone layer had slowed down due to the ban on CFCs. Scientists have developed HCFC to replace CFC for the same purpose. These HCFCs are short-lived in the atmosphere to reach the stratosphere and damage ozone layer.

Ozone Layer Depletion by Nitric Oxide

Chemistry of ozone depletion by nitric oxide is shown below:

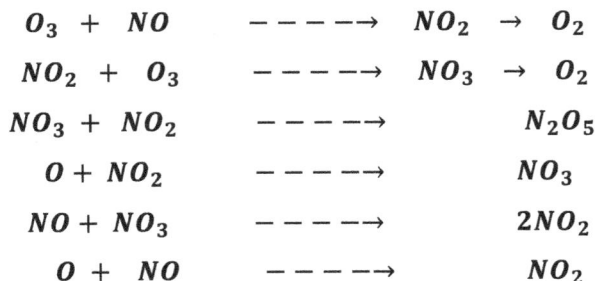

$$O_3 + NO \quad ----\rightarrow \quad NO_2 \rightarrow O_2$$
$$NO_2 + O_3 \quad ----\rightarrow \quad NO_3 \rightarrow O_2$$
$$NO_3 + NO_2 \quad ----\rightarrow \quad N_2O_5$$
$$O + NO_2 \quad ----\rightarrow \quad NO_3$$
$$NO + NO_3 \quad ----\rightarrow \quad 2NO_2$$
$$O + NO \quad ----\rightarrow \quad NO_2$$

When a nitric oxide (NO) molecule combines with O3, it is oxidized to nitrogen dioxide (NO2). This NO2 then combines with another O3 molecule to become NO3 and O2. Both NO2 and NO3 may combine to form N2O5. Even if atomic oxygen is available, it readily combines with NO2 to yield NO3. Thus, O3 is completely utilized for the above reactions and thereby depleted. International community, after realizing its seriousness, has agreed to withdraw the operation of jet airplanes that emit NO in stratosphere.

Fossil Fuels and their Impact on the Environment:

The technical definition of fossil fuels is "incompletely oxidized and decayed animal and vegetable materials, specifically coal, peat, lignite, petroleum and natural gas" or "material that can be burned or otherwise consumed to produce heat". In our modernized western world, fossil fuels provide vast luxurious importance. We retrieve these fossil fuels from the ground and under the sea and convert them into electricity. Approximately 90% of the world's electricity demand is generated from the use of fossil fuels.

There is a growing concern regarding the collaboration between fossil fuels and environmental pollution. Debates regarding this contamination have become commonplace in today's effort to sustain the earth's health. Fossil fuels are not considered a renewable energy source and aside from the environmental impact, the cost of retrieving and converting them is beginning to demand notice. Seemingly this issue has many different angles that need to be addressed in order to ensure future generations a sustainable living.

Combustion of these fossil fuels is considered to be the largest contributing factor to the release of greenhouse gases into the atmosphere. In fact it is believed that energy providers are the largest source of atmospheric pollution today. There are many types of harmful outcomes which result from the process of converting fossil fuels to energy. Some of these include air pollution, water pollution, accumulation of solid waste, not to mention the land degradation and human illness.

Evidence of the ill effects of fossil fuels is endless, and can take on many forms. Some forms are not easily seen by the human eye, although the disastrous results such as the loss of aquatic life can be seen somewhat after the fact. Carbon dioxide is considered the most prominent contributor to the global warming issue. The impact of global warming on the environment is extensive and affects many areas. In the Antarctica, warmer temperatures may result in more rapid ice melting which increases sea level and compromises the composition of surrounding waters. Rising sea levels alone can impede processes ranging from settlement, agriculture and fishing both commercially and recreationally.

Air pollution is another problem arising from the use of fossil fuels, and can result in the formation of smog causing human problems and affecting the sustainability of crops. Smog seeps through the protective layer on the leaves and destroys essential cell membranes. This results in smaller yields and weaker crops, as the plants are forced to focus on internal repair and do not thrive. Many toxic substances are released during the conversion or retrieval process including "Vanadium" and "Mercury". According to the "New Book of Popular Science", "it is suspected that significant quantities of Vanadium in the atmosphere results from residual fuel oil combustion".

When coal is burned, it releases nitrous oxide. Unfortunately this is kept in the atmosphere for very long time. The harmful impact of this chemical could take up to a couple of hundred years to make itself known. It is very difficult to prevent or to diminish an impact when you are not even aware of what it may be. The only solution in this case is to reduce the formation of nitrous oxide. Nearly 50% of the nitrogen oxide and 70% of sulfur dioxide in the atmosphere are direct result of emissions released when coal is burned.

Converting fossil fuels may also result in the accumulation of solid waste. This type of accumulation has a devastating impact on the environment. Waste requires adequate land space for containment and/or treatment, as well as financial support and monitoring for wastes not easily disposed of. This type of waste also increases the risk of toxic runoff which can poison surface and groundwater sources for many miles. Toxic runoff also endangers surrounding vegetation, wildlife, and marine life.

Delivery of fossil fuels can result in oil spills, and many of us are familiar with the impacts of this type of disaster. Seepage from foundations like that of oil rigs and pipelines can also result in similar destruction for habitat and wildlife. According to the Department Of The Interior, vast damage to waterways can be attributed to the extraction of coal. Coal extraction may very well be the leading the source of water pollution today.

Use of unleaded gas has helped to reduce the release of lead into the environment. Although in third world countries, the safer unleaded gas has not been fully utilized and is still a major concern. Unfortunately for developing countries, the economy and technology available to them is quite behind what we are used to. With this in mind many environmental issues are treated at an international level, which allows for more efficient handling.

We have become a very energy greedy generation and our demands for electricity are very high. As far as reducing these harmful effects, we must first reduce our demand. Science may be able to find alternative, healthier sources, although not ones that meet the required supply. These types of horrendous impacts are felt globally and should not be considered one countries problem. Sometimes social limitations and/or economic stability can make the process of change very difficult.

One thing is for sure, that by being more energy efficient and conservative, we will be helping to alleviate the toll on environmental and human health.

LEGISLATION:

The main cause of environmental degradation is the human activity in one way or the other. Law (legislation) is the only regulator of these activities. So in environmental conscious states the environmental problems are generally handled at the legislative level. In India various laws for the protection of environment have been enacted from time to time and is perhaps the first constitution in the world to contain specific provisions for the protection and improvement of environment. In view of various constitutional provisions and other statutory provisions contained in various laws relating to environmental protection, the Supreme Court has held that the essential features of *"Sustainable Development"* such as the *"precautionary principle"* and the *"polluter pays principle"* are part of the environment law of the country. The constitution of India also obligates the *"state"* as well as *"citizen"* to protect and improve the environment. The legislative measures have been followed by other nations of the world as well e.g. the framers of the constitution of South Africa were greatly influenced by the provisions relating environmental protection and they incorporated similar provisions in their constitution. Likewise other countries of the world have framed different types of legislations regarding the environmental protection. Soon after the world famous Stockholm Conference of 1972, India took substantial measures for environmental protection and passed different types of Acts relating to different components of environment as:

- Wild life protection Act, 1972

- Forest conservation Act, 1980 (Amended in 1988)

- The water (prevention and control of pollution) Act, 1974 amended in 1988

- The Air (prevention and control of pollution) Act, 1981 amended in 1987

- The Environmental protection Act, 1986

The Air (Prevention and Control of Pollution) Act, 1981:

This act was passed under article 253 of Indian constitution in pursuance to the decisions of Stockholm conference. The objective of this act is to provide for the prevention, control and abatement of air pollution in order to preserve the quality of air.

The Act defines relevant terms such as air pollution, pollutant, automobile, industrial plant, control equipment, hazardous substances etc. The Act provides the declaration of certain heavily polluted areas as "Air pollution control areas" where no industrial plant shall be operated without prior consent of State Pollution Control Boards (SPCB's). The central and state boards have been entrusted with the task of controlling and preventing pollution and accordingly they have been re-designated as CPCB and SPCB's respectively. The state boards have been empowered to lay down and enforce standards for prevention and control of air pollution. The state government in consultation with the respective board may give instructions to the concerned authority of registrations under the motor vehicles Act 1939 to ensure emission standards from automobiles. The state boards have the powers to sue a polluter in a court of law to prevent him from polluting the environment (air) and the expenses incurred by the boards for doing so will be recovered from the polluter. Further the boards have powers to authorize any person to enter and inspect the premises of the polluter and to collect the samples of emission for the analysis of pollutants. Failure to comply with the conditions prescribed for this purpose is punishable with a fine and imprisonment. The Act was amended in 1987 to render it more effective and to include "noise" also under the definition of air pollutants.

Dilemma of Developing Countries

Development -the dream of the developing countries has brought about a marked deterioration in some of their environmental conditions. In these countries development is carried out but at the cost of environment which has thus become a cause of environmental deterioration. The record figure of 70% population living in worlds developing countries determines the extent and intensity of exploitation of nature and its resource base. The resource exploitation in these countries is also determined by their political, social and cultural institution. The major causes of environmental deterioration in the developing countries are over exploitation, over population and poverty.

In order to feed and sustain millions of poor people, the tropical region of the world colonized by the developing countries and underdeveloped countries (mega-diversity regions) endowed with rich and diverse flora and fauna, are forced to expand agriculture and industries resulting in over-exploitation of resources and large scale deforestation. The forest destruction of the tropical ecosystems is proving disastrous in the entire biosphere.

Destruction of forest in the third world countries is established to be more than 11 million hectares every year. It is because of subsequent conversion of forests to pastures, shifting cultivation and timber extraction which leads to loss of biodiversity and forest and impoverishment of rural masses. Over-grazing leading destruction of protective vegetation cover is also a common problem of the drought affective parts of the developing countries such as Subsharan Africa. Deforestation, largely for logging and fuel wood extraction by the developing countries leading to soil erosion is also a problem observed there. A number of socio-economic and political factors have also been associated with accelerated soil erosion. These include population pressure, skewed land resource distribution, poverty, marginalization, and increasing demand of fuel wood etc.

In many developing countries, population growth is rapid and the demand for agricultural land and fuel wood is ever increasing. Furthermore, agricultural systems are characterized by skewed land distribution, land owners control on majority of the land. The poor and weak farmers are thus forced into marginal land, which is very much susceptible to erosion. Generally viewed in developing countries is:

1. Forests are cut down for economic and commercial beginnings.

2. Population growth reduces per capita land use.

3. Depletion of water resources, increased energy consumption and faulty agricultural system.

Lack of Economic Capacity to Deal with Atmospheric Pollution:

Economic base of a country is directly related to the comforts and problems faced by its citizens. Stronger the economy, stronger is the defense mechanism of the country to combat the problems. Weaker the economy, weaker is the defense mechanism of the country to combat the problems. Secure economy plays a very significant role in tackling almost each and every type of problem. However, lack of economic capacity/insecure economy intensifies the problems (atmospheric

pollution in this case) because abatement of atmospheric pollution needs the adoption of following strategies which indeed is impossible for countries with insecure economy.

i) Research and developing programmes: Leading institutes to undertake specific projects (needing lot of funds) for seeking solution to the problem.

ii) Monitoring programmes: For monitory the environment in an area, new techniques like Remote Sensing and Geographic Information System (GIS) need to be used out. But the use of these technologies requires lots and lots of funds which is beyond the potential of the countries with insecure economies.

iii) Technical personals: For analytical monitoring and abatement jobs, there is a need of designing new instruments that would give accurate results so as to control a problem in an efficient way. Besides, this lack of economic capacity leads to poverty which leads to inequalities and discriminations, which in turn leads to activities like deforestation and terrestrial congestion, thus disturbing the natural atmospheric balance.

Role of Government (Local and National):

Ever since man evolved from its ancestors, the struggle to win nature has ensued. He has made and is continuing to make every conscious attempt to beat nature into submission but is in turn finding himself entangled in a number of problems that haunt him like a ghost. A handful of these problems include population explosion, rampant urbanization, shrinking forest, expanding deserts, loss of biodiversity, air pollution, and food security. To tackle these deadly environmental problems, the Government (both local and national) has a very important role to play. The role that a government can play is:

1) Framing of strict laws and their proper implementation.

2) Issuing of guidelines for different activities having impacts on environment.

3) Conducting EIA (Environmental Impact Assessment) of different projects.

4) Implementation of more stringent environment measures and national environmental policy.

5) Preparation of more comprehensive guide lines for EIA.

6) Fiscal incentives to enterprises and organization devoted to environmental protection should be given and expenses on environmental protection should become tax deducible to encourage their efforts.

7) Additional tax liability on companies emitting polluting gases like SO2 and discharging liquid effluents beyond permissible limits.

8) For integration of environment protection, the ministry of environment in the government should be ranked high which typically ranks low in the cabinet hierarchy.

9) Govt. should place the issues of environmental awareness and public participation on top priority. Provisions of information and institutional arrangements should be there.

10) Special funding policies and programmes to encourage the introduction of cleaner and low/no-waste technologies.

11) Govt. should ban the use of environmentally dangerous products.

12) Govt. should take effective steps for development of environmentally friendly products and to avoid publicity of non-sustainable development.

13) Govt. should take effective steps to implement the international agreement in their respective nations.

Responsibility of Industry:

Industries have a great role to pray in the field of environmental protection. A few of the responsibilities of an industry are.

1. Industry should have a proper environmental policy and institute environmental policy audit as an integral part of their corporate responsibility.

2. It should concentrate on the use of new and renewable sources of energy such as solar energy, wind energy and bio-energy.

3. It should emphasize on 3R approach.

4. There should be environmental rethinking by the industry on how its activities impact the environment.

5. Industry should look beyond the short term benefits.

6. Industry should include in its balance sheet an assessment of environmental damage for which its shareholders will be liable in future.

7. Manufactures should mark environment friendly products with a special logo (eco-labeling) which will empower consumers to make decisions based on environmental impact.

8. Industry should respect the national laws and international agreements related to environment.

9. Industry should respect the emission standard laid down by various agencies.

10. Industry should also make use of air pollution control devices.

Role of Environmental Organization:

The environmental organizations which are a product of social action, history and culture have emerged a new sector and can play an important vis-à-vis environmental protection. A few roles that the organizations can play are as follows:

1. They are actively involved in major developmental activities and in the promotion of environmental protection.

2. They can help in exposing the indifferences shown by Govt. to various vital environmental issues.

3. Help in bringing the administration and the masses together for the implementation and execution of progressive and vital management plans.

4. They can cultivate a political culture where organizations work in co-operation and not in rivalry for environmental protection.

5. They can educate the citizens, work as catalysts, and disseminate information, shape public opinion by bringing awareness among people and initiate proposals for the development of masses.

6. They can help in the proper execution of the plans of the Govt. and also convey the policies to the public in convincing manner.

7. They can play the role by focusing on the vital issues and ensuring action at local level.

8. They can make pressure groups in their respective areas and get the task done.

9. They help in the promotion of innovative chemical technologies to reduce or eliminate the use or generation of hazardous waste in design, manufacture and use.

10. To help people to acquire necessary knowledge about environmental problems.

11. To develop necessary skills to solve the environmental problems.

12. To create a conducive atmosphere to make it possible for the citizens to participate in decision make where and when it concerns for environment.

13. To provide awareness of economic, political and ecological interdependence.

14. To propagate knowledge about reusing of scrap and excess material.

15. To propagate knowledge about using recyclable resource.

16. To show the possible ways of selecting environmentally compatible materials, use of less energy resources and minimization of wastes in production process.

17. Successful implementation of environmental protection programs depends on peoples participation along with the involvement of environmental organization both Govt. and Non-Govt. organizations.

Ozone Layer -The Protection Shield:

Ozone is a naturally occurring pungent smelling, bluish gas occurring throughout the atmosphere with a maximum mixing ration at altitudes ranging from 15-35 kms above earth. This region is frequently known as ozone layer. The ozone concentration differs by about 10ppm in the stratosphere as compared to 0.05ppm in troposphere. Near the earth, the elevated level of ozone can be toxic to both plants and animals. But it acts as a protective cover for life on earth when present in stratosphere. It strongly absorbs the incoming UV-radiation from sun in the wavelength range of 220-330 nm and thereby protects the life on earth from severe radiation damage such as DNA mutation, skin cancer and eye damage etc. In absence of this layer, all the UV-rays of the sun will reach the earth's surface and consequently the temperature of the lower atmosphere will rise to such an extent that the biological furnace of biosphere will turn into blast furnace.

The thickness of O3 layer is measured in Dobson units (DU) with 1 DU equivalent to 0.01 mmHg of compressed gas at 0°C and 760 mmHg of pressure. The average thickness of O3 layer in stratosphere has been estimated to be about 250 DU. The level above 250 Dobson units are considered as normal but when its level decreases below 250 DU the condition generally qualifies as ozone-hole. The term ozone hole is actually used to describe the thinning of ozone layer.

Formation of Ozone: Ozone is formed in the stratosphere under natural conditions by the following photochemical reactions.

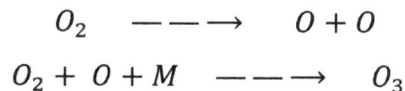

$$O_2 \quad ---\rightarrow \quad O + O$$
$$O_2 + O + M \quad ---\rightarrow \quad O_3$$

Where M is the third body which absorbs the excess energy liberated by the above reaction and thereby stabilizes the ozone molecule.

In the lower mesosphere, the atmospheric oxygen absorb UV-radiations (>240nm) and photo dissociates it into two oxygen atoms. These oxygen atoms subsequently react with molecular oxygen of upper stratosphere and produce ozone. It is also formed at ground level in a very little concentration of about 0.05 ppm where it acts as a harmful pollutant for both plants and animals. Nitrogenous gases react with the UV-radiations from the sun and result in the formation of ozone as follows.

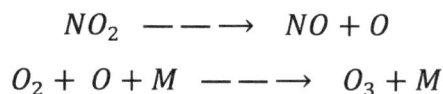

$$NO_2 \quad ---\rightarrow \quad NO + O$$
$$O_2 + O + M \quad ---\rightarrow \quad O_3 + M$$

Mechanism of Ozone Depletion: The first conscious effort about ozone depletion was made by

M. Molino and S. Rowland, two professors of University of California in 1974 -1975. The British Antarctic survey team in 1835 and the multinational expedition of Antarctica in 1987 confirmed the prediction about O3 depletion. They also provided the information about the creation of ozone hole over Antarctica. The problem of ozone depletion and its adverse effects have threatened the existence of life on earth. It is a global problem and is caused by.

i)   Natural Processes and

ii) Anthropogenic Processes

i) Natural Process: Under natural conditions the ozone gets depleted by the collision of O_3 with monatomic oxygen in the following manner.

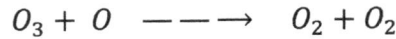

$$O_3 + O \quad \text{---} \rightarrow \quad O_2 + O_2$$

OR

$$O_3 + O \quad \text{---} \rightarrow \quad 2O_2$$

This natural phenomenon of ozone depletion however does not necessarily upset the ozone equilibrium because the depletion of ozone is often compensated by the formation of ozone through atmospheric circulation.

ii) Anthropogenic Processes: Anthropogenic processes like industrialization, automobile emission, nuclear explosion etc. are also responsible for the depletion of ozone because these activities release a large number of pollutants with O_3 depletion potential. These substance e.g. Chloroflouro carbons (CFC's), Oxides of Nitrogen (NOx), Hydroxyl radicals (OH) are called as ozone depleting substances (ODS). Depletion of O_3 by these substances take place in the following manner.

a)   By Hydroxyl radical (OH): The hydroxyl radical is generated by either of the two photochemical reactions (i) and (ii).

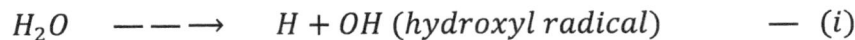

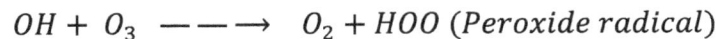

$$H_2O \quad \text{---} \rightarrow \quad H + OH \; (hydroxyl \; radical) \qquad - (i)$$
$$H_2O + O \quad \text{---} \rightarrow \quad 2OH \qquad\qquad\qquad - (ii)$$
$$OH + O_3 \quad \text{---} \rightarrow \quad O_2 + HOO \; (Peroxide \; radical)$$
$$HOO + O \quad \text{---} \rightarrow \quad OH + O_2$$
$$H + O_3 \quad \text{---} \rightarrow \quad O_2 + OH$$

c)   By Nitric Oxide (NO): The nitric oxide is generated by either of the reactions (i) and (ii)

$$N_2O + O \quad \text{---} \rightarrow \quad 2NO \qquad - (i)$$
$$NO_2 \quad \text{---} \rightarrow \quad NO + O \qquad - (ii)$$
$$NO + O_3 \quad \text{---} \rightarrow \quad NO_2 + O_2$$
$$NO_2 + O \quad \text{---} \rightarrow \quad O_2 + NO$$

c) By Oxides of Nitrogen (NOx): Produced by nuclear explosion and introduced directly into the stratosphere.

$$NO_2 + O_3 \text{---} \rightarrow NO_3 + O_2$$

c) By Chlorine Radical: The CFC's used as coolants in refrigerators, air conditioners, as propellants in aerosol sprays and in plastic foams such as "Thermo Cole" or "Styrofoam" are the most important destroyers of ozone layer. The CFC molecules, escaping into the atmosphere, decompose to release chlorine radical (by photo dissociation), the chlorine radical released reacts again and again and hence results in the loss of thousands of ozone molecules. The release of chlorine radical and the depletion of ozone by it take place in the following way.

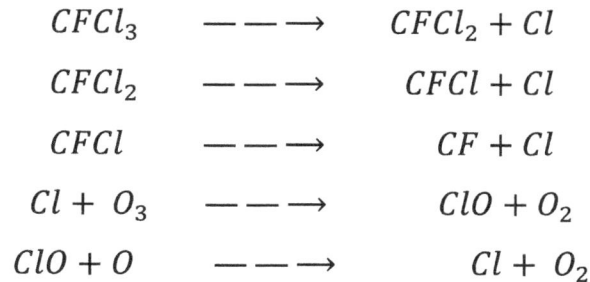

$$CFCl_3 \quad ---\!\!\rightarrow \quad CFCl_2 + Cl$$

$$CFCl_2 \quad ---\!\!\rightarrow \quad CFCl + Cl$$

$$CFCl \quad ---\!\!\rightarrow \quad CF + Cl$$

$$Cl + O_3 \quad ---\!\!\rightarrow \quad ClO + O_2$$

$$ClO + O \quad ---\!\!\rightarrow \quad Cl + O_2$$

Effects of Ozone Depletion: As the ozone layer acts as a protective umbrella in the stratosphere, its depletion has serious consequences on the whole universe and some of them are as follows.

i) Effects on climate: Because of weakening of the O3 layer, there is less absorption of UV radiations and consequent rise in temperature. The substantial rise in earth's temperature would cause global warming and climate changes at both regional and local level.

ii) Effects on human beings:

a) Exposures to UV radiations cause skin cancer of three types viz. basal cell carcinoma, squamous cell carcinoma and melanoma particularly among the white population.

b) Lagerhans cells in the epidermis of human skin are the key players in the immune surveillance. UV radiation get them damaged first, thus decrease the defence mechanism of the skin.

c) Damages the eyes -especially leads to the development of cataracts.

d) UV radiations cause blood vessels near the skin surface to carry more blood, making the skin hot, swollen and red hence causing sun burns.

e) UV radiations cause leukemia and breast cancer although the reasons are not clear.

f) Suppress the body immune system (responses), thus making the human body more prone to infectious diseases.

g) UV exposure causes mutations in the human beings.

iii) Effects on Plants:

a) UV radiations cause injuries to plant proteins as they are susceptible to UV injury.

b) Chlorophyll reduction and harmful mutation are also observed in plants.

c) Intense UVradiation exposure causes greater evaporation.

d) Typically sensitive plants show reduced growth and smaller leaves.

iv) Effects on Ecosystem: The effects on the aquatic ecosystem mainly depend upon the depth to which the UV radiations penetrate. Enhanced UV radiation exposures have shown their serious effect on the range of small organism like zooplanktons, larval crabs, shrimps and juvenile fish as well as slowing rate of photosynthesis in phyto-planktons.

International Efforts to Combat Ozone Depletion:

Vienna Convention: This was the first convention on the protection of ozone layer. It was held in March 1985 and was the first and foremost convention to highlight the problems related to ozone and immediate necessity on the part of the international community to take steps to protect the lifesaving umbrella from further damage. In spite of the fact that the convention contained no substantive provisions for the protection of ozone layer and controlling of the substance having ozone depleting potential, it created a frame work for the much stronger international initiative called Montreal Protocol.

Montreal Protocol: This protocol was held on 16 September, 1987 and came into force on 1st January 1989. This protocol was held with an aim which called for a complete ban on the manufacture of CFC's, the substance with strongest ozone depletion potential. This ban was to take effect from 1989. Government representatives of 150 countries signed this agreement which Planned phasing out of ozone destroying chemicals and made commitment with themselves to protect the O3 layer from further damage. It helped the participatory countries to corporate with each other in scientific research to improve understanding of the atmospheric processes and to reduce and eventually eliminate the emissions of man-made O3 depleting substances (ODS's). India signed this Protocol in 1992. It contains clear provisions which impose obligations in states to:

i.   Reduce production and consumption of ODS's.

ii. Co-operate in developing alternate substances.

iii. Restrict trade in these substances.

iv. Follow innovative approaches to the issue of enforcement.

v.   Adding incentives for countries joining Montreal protocol.

The protocol called for a gradual reduction of both consumption and production of the ozone depleting substances over an eight year period with 1986 as the base level. The overall effect was to freeze the consumption of these substances. The major highlights of the protocol were:

•   CFC's to be phased out by the year 2000 with cuts of 50% by 1995 and 85% by 1997.

•   Halones to be phased out by 2000, with an 85% cut by 1995.

•   Methyl chloroform to be phased out by 2005, with a 30% cut by 2000.

The developing countries demanded positive and definite assurance of compensation and transfer of technology from industrialized nations. They justified their demand on the basis of "Polluter Pays" and not "Victim Pays principle". For this developing countries were given a 10 year "grace period" to phase out of the CFC's. It also included innovative finding provisions for less effluent member countries so that financial and technical incentives (such as a transfer of technology) could be given to them.

It also contains provisions to deal with the problems of few nations that have not signed the protocol and continue to produce and consume CFC's by banning trade in these substances with non-member states. Thus, parties to the protocol are prohibited from importing such substances or exporting CFC production technology and equipment.

6

Water Pollution

Water pollution is a large set of adverse effects upon water bodies (lakes, rivers, oceans, and groundwater), caused by human activities. Although natural phenomena such as volcanoes, storms, earthquakes etc. also cause major changes in water quality and the ecological status of water bodies, these are not deemed to be pollution. Water pollution has many causes and characteristics. Increase in nutrient loading may lead to eutrophication. Organic wastes such as sewage and farm waste impose high oxygen demands on the receiving water leading to oxygen depletion with potentially severe impacts on the whole eco-system. Industries discharge a variety of pollutants in their wastewater including heavy metals, organic toxins, oils, nutrients, and solids. Discharges can also have thermal effects, especially those from power stations, and these too reduce the dissolved oxygen. Silt-bearing runoff from many activities including construction sites, deforestation and agriculture can inhibit the penetration of sunlight through the water column, restricting photosynthesis and causing blanketing of the lake or river bed, in turn damaging ecological systems. Pollutants in water include a wide spectrum of chemicals, pathogens, and physical chemistry or sensory changes. Many of the chemical substances are toxic or even carcinogenic. Pathogens can obviously produce waterborne diseases in either human or animal hosts. Alterations of water's physical chemistry include acidity, conductivity, temperature, and eutrophication. Eutrophication is the fertilization of surface water by nutrients that were previously scarce. Even many of the municipal water supplies in developed countries can present health risks.

Sources of Water Pollution:

Principal sources of water pollution are:

- Industrial discharge of chemical wastes and byproducts.

- Discharge of poorly-treated or untreated sewage.

- Surface runoff containing pesticides.

- Slash and burn farming practice, which is often an element within shifting cultivation agricultural systems.

- Surface runoff containing spilled petroleum products.

- Surface runoff from construction sites, farms, or paved and other impervious surfaces e.g. silt.

- Discharge of contaminated and/or heated water used for industrial processes.

- Acid rain caused by industrial discharge of sulfur dioxide (by burning high sulfur fossil fuels).

- Excess nutrients added by runoff containing detergents or fertilizers.

- Underground storage tank leakage, leading to soil contamination and hence aquifer contamination.

Groundwater Pollution: Groundwater pollution is a type of pollution which occurs when groundwater becomes contaminated. It is an introduction of certain pollutant(s) into the groundwater which reduces the quality of groundwater making its use very limited, or in some cases impossible. Many different chemicals and various synthetic products we use today are usually the main causes of groundwater pollution.

Groundwater pollution is a very serious problem. Unlike a lot of the pollution on the surface waters, like trash floating in the bay, groundwater pollution is harder to recognize until after illness has occurred. Around the world, groundwater pollution is a very serious and costly problem. Once contaminated, groundwater is very expensive to clean up and make usable again, and in some cases, an aquifer may be so contaminated that it has to be abandoned, which can put tremendous pressure on a community as it attempts to find a new supply of water. The risk of groundwater pollution is increasing both from the disposal of waste materials and from the widespread use by industry and agriculture of potentially polluting chemicals in the environment. Groundwater pollution can occur either as discrete, point sources (e.g. from landfills), or from the wider, more diffuse use of chemicals, such as the application to and fertilizers and pesticides and the deposition of airborne pollutants in heavily industrialized regions. Changes in groundwater quality may result from direct or indirect anthropogenic activities. Direct influence occurs as a result of the introduction of natural or artificial substances derived from human activities into groundwater. Indirect influences are those changes in groundwater quality caused by human interference with hydrological, physical and biochemical processes, but without the addition of substances. The main contaminants of groundwater are heavy metals, organic chemicals, fertilizers, bacteria and viruses. The enormous range of contaminants encountered in groundwater reflects the wide range of human economic activities in the world. The major activities generating contaminants are associated with agricultural, mining, industrial and domestic sectors.

Domestic Sewage and Its Treatment:

Sewage treatment, or domestic wastewater treatment, is the process of removing contaminants from wastewater. It includes physical, chemical and biological processes to remove physical, chemical and biological contaminants. Its objective is to produce a waste stream suitable for discharge back into the environment. This material is often inadvertently contaminated with toxic organic and inorganic compounds.

Sewage is created by residential, institutional, commercial and industrial establishments. It can be treated close to where it is created (in septic tanks, biofilters or aerobic treatment systems), or collected and transported via a network of pipes and pump stations to a municipal treatment plant for treatment. Industrial sources of wastewater often require specialized treatment processes. Raw influent (sewage) is the liquid waste from toilets, bathrooms, showers, kitchens, sinks etc. Municipal wastewater therefore includes residential, commercial, and industrial liquid waste discharges, and may include stormwater runoff. Sewage systems capable of handling stormwater are known as combined systems.

Typically, sewage treatment involves three stages, called *primary, secondary and tertiary* treatment. First, the solids are separated from the wastewater stream. Then dissolved biological matter is progressively converted into a solid mass by using indigenous, water-borne bacteria. Finally, the biological solids are neutralized then disposed of or re-used, and the treated water may be disinfected chemically or physically (for example by lagoons and micro-filtration). The final effluent can be discharged into a stream, river, bay, lagoon or wetland, or it can be used for the irrigation of a golf course, green way or park. If it is sufficiently clean, it can also be used for groundwater recharge.

Sewage Treatment Methods:

The site where the raw wastewater is processed before it is discharged back to the environment is called a wastewater treatment plant (WWTP). The order and types of mechanical, chemical and biological systems that comprise the wastewater treatment plant are typically the same in most of the countries:

Mechanical treatment;

- Influx (Influent)

- Removal of large objects

- Removal of sand and grit

- Pre-precipitation

Biological treatment;

- Oxidation bed (oxidizing bed) or aeration system

- Post precipitation

- Effluent

Chemical treatment (this step is usually combined with settling and other processes to remove solids, such as filtration.)

Primary Treatment:

Primary treatment removes the materials that can be easily collected from the raw wastewater and disposed off. The typical materials that are removed during primary treatment include fats, oils, and greases (also referred to as FOG), sand, gravels and rocks (also referred to as grit), larger settleable solids including human waste and floating materials. This step is done entirely with machinery, hence the name mechanical treatment.

Influx (Influent) and Removal Of Large Objects

In the mechanical treatment, the influx (influent) of sewage water is strained to remove all large objects that are deposited in the sewer system, such as rags, sticks, condoms, sanitary towels (sanitary napkins) or tampons, cans, fruit, etc. This is most commonly done with a manual or automated mechanical screen. This type of waste is removed because it can damage or clog the equipment in the sewage treatment plant.

Sand and Grit Removal

Primary treatment typically includes a sand or grit channel or chamber where the velocity of the incoming wastewater is carefully controlled to allow sand, grit and stones to settle, while keeping the majority of the suspended organic material in the water column. This equipment is called a detritor or sand catcher. Sand grit and stones need to be removed early in the process to avoid damage to pumps and other equipment in the remaining treatment stages. Sometimes there is a sand washer (grit classifier) followed by a conveyor that transports the sand to a container for disposal. The contents from the sand catcher may be fed into the incinerator in a sludge processing plant, but in many cases, the sand and grit is sent to a landfill.

Sedimentation

Many plants have a sedimentation stage where the sewage is allowed to pass slowly through large tanks, commonly called "primary clarifiers" or "primary sedimentation tanks". The tanks are large enough that suspended solids can settle and floating material such as grease and oils can rise to the surface and be skimmed off. The main purpose of the primary stage is to produce a generally homogeneous liquid capable of being treated biologically and a sludge that can be separately treated or processed. Primary settlement tanks are usually equipped with mechanically driven scrapers that continually drive the collected sludge towards a hopper in the base of the tank from where it can be pumped to further sludge treatment stages.

Secondary Treatment

Secondary treatment is designed to substantially degrade the biological content of the sewage such as are derived from human waste, food waste, soaps and detergent. The majority of municipal and industrial plants treat the settled sewage liquor using aerobic biological processes. For this to be effective, the biota requires both oxygen and a substrate on which to live. There are number of ways in which this is done. In all these methods, the bacteria and protozoa consume biodegradable soluble organic contaminants (e.g. sugars, fats, organic short-chain carbon molecules, etc.) and bind much of the less soluble fractions into flocs. Secondary treatment systems are classified as fixed film or suspended growth. Fixed-film treatment process including trickling filter and rotating biological contactors where the biomass grows on media and the sewage passes over its surface. In suspended growth systems—such as activated sludge—the biomass is well mixed with the sewage and can be operated in a smaller space than fixed-film systems that treat the same amount of water. However, fixed-film systems are more able to cope with drastic changes in the amount of biological material and can provide higher removal rates for organic material and suspended solids than suspended growth systems.

Roughing Filters

These are intended to treat particularly strong or variable organic loads, typically industrial, to allow them to then be treated by conventional secondary treatment processes. Characteristics include typically tall, circular filters filled with open synthetic filter media to which wastewater is applied at a relatively high rate. On larger installations, air is forced through the media using blowers. The resultant wastewater is usually within the normal range for conventional treatment processes.

Activated Sludge

Activated sludge plants encompasses a variety of mechanisms and processes that uses dissolved oxygen to promote the growth of biological flocs that substantially remove organic material. It also traps particulate material and can, under ideal conditions, convert ammonia to nitrite and nitrate ultimately to nitrogen gas.

Fluidized Bed Reactors

The carbon adsorption following biological treatment was particularly effective in reducing both the BOD and COD to low levels. A fluidized bed reactor is a combination of the most common stirring tank, packed bed, continuous flow reactors. It is very important to chemical engineering because of its excellent heat and mass transfer characteristics. In a fluidized bed reactor, the substrate is passed upward through the immobilized enzyme bed at a high velocity to lift the particles. However the velocity must not be so high that the enzymes are swept away from the reactor entirely. This causes low mixing; these types of reactors are highly suitable for the exothermic reactions. It is most often applied in immobilized enzyme catalysis.

Filter Beds (Oxidizing Beds)

In older plants and plants receiving more variable loads, trickling filter beds are used where the settled sewage liquor is spread onto the surface of a deep bed made up of coke (carbonised coal), limestone chips or specially fabricated plastic media. Such media must have high surface areas to support the biofilms. The liquor is distributed through perforated rotating arms radiating from a central pivot. The distributed liquor trickles through this bed and is collected in drains at the base. These drains also provide a source of air which percolates up through the bed, keeping it aerobic. Biological films of bacteria, protozoa and fungi form on the media's surfaces and eat or otherwise reduce the organic content. This biofilm is grazed by insect larvae and worms which help maintain an optimal thickness. Overloading of beds increases the thickness of the film leading to clogging of the filter media and ponding on the surface.

Biological Aerated Filters

Biological Aerated (or Anoxic) Filter (BAF) or Biofilters combine filtration with biological carbon reduction, nitrification or denitrification. BAF usually includes a reactor filled with a filter media. The media is either in suspension or supported by a gravel layer at the foot of the filter. The dual purpose of this media is to support highly active biomass that is attached to it and to filter suspended solids. Carbon reduction and ammonia conversion occurs in aerobic mode and sometime achieved in a single reactor while nitrate conversion occurs in anoxic mode. BAF is operated either in upflow or downflow configuration depending on design specified by manufacturer.

Membrane Biological Reactors

Membrane biological reactors (MBR) combines activated sludge treatment with a membrane liquid-solid separation process. The membrane component utilizes low pressure microfiltration or ultra filtration membranes and eliminates the need for clarification and tertiary filtration. The membranes are typically immersed in the aeration tank (however, some applications utilize a separate membrane tank). One of the key benefits of a membrane bioreactor system is that it effectively overcomes the limitations associated with poor settling of sludge in conventional activated sludge (CAS) processes. The technology permits bioreactor operation with considerably higher mixed liquor suspended solids (MLSS) concentration than CAS systems, which are limited by sludge settling. The process is typically operated at MLSS in the range of 8,000–12,000 mg/L, while CAS is operated in the range of 2,000–3,000 mg/L. The elevated biomass concentration in the membrane bioreactor process allows for very effective removal of both soluble and particulate biodegradable materials at higher loading rates. Thus increased Sludge Retention Times (SRTs)—usually exceeding 15 days—ensure complete nitrification even under extreme cold weather operating conditions. The cost of building and operating a MBR is usually higher than conventional wastewater treatment, however, as the technology has become increasingly popular and has gained wider acceptance throughout the industry, the life-cycle costs have been steadily decreasing.

Secondary Sedimentation

The final step in the secondary treatment stage is to settle out the biological floc or filter material and produce sewage water containing very low levels of organic material and suspended matter.

Rotating Biological Contactors

Rotating biological contactors (RBCs) are mechanical secondary treatment systems, which are robust and capable of withstanding surges in organic load. The rotating disks support the growth of bacteria and micro-organisms present in the sewage, which breakdown and stabilize organic pollutants. To be successful, micro-organisms need both oxygen to live and food to grow. Oxygen is obtained from the atmosphere as the disks rotate. As the micro-organisms grow, they build up on the media until they are sloughed off due to shear forces provided by the rotating discs in the sewage. Effluent from the RBC is then passed through final clarifiers where the microorganisms in suspension settle as sludge. The sludge is withdrawn from the clarifier for further treatment.

Tertiary Treatment

Tertiary treatment provides a final stage to raise the effluent quality before it is discharged to the receiving environment (sea, river, lake, ground, etc.). More than one tertiary treatment process may be used at any treatment plant. Removal of additional BOD and TSS from secondary effluent is referred as effluent polishing.

Filtration

Sand filtration removes much of the residual suspended matter. Filtration over activated carbon removes residual toxins.

Lagooning

Lagooning provides settlement and further biological improvement through storage in large man-made ponds or lagoons. These lagoons are highly aerobic and colonization by native macrophytes, especially reeds, is often encouraged. Small filter feeding invertebrates such as Daphnia and species of Rotifera greatly assist in treatment by removing fine particulates.

Constructed Wetlands

Constructed wetlands include engineered reedbeds and a range of similar methodologies, all of which provide a high degree of aerobic biological improvement and can often be used instead of secondary treatment for small communities.

Waste Removal

Wastewater may contain high levels of the nutrients nitrogen and phosphorus. Excessive release to the environment can lead to a buildup of nutrients, called eutrophication, which can in turn encourage the overgrowth of weeds, algae, and cyanobacteria (blue-green algae). This may cause an algal bloom, a rapid growth in the population of algae. The algae numbers are unsustainable and eventually most of them die. The decomposition of the algae by bacteria uses up so much of oxygen in the water that most or all of the animals die, which creates more organic matter for the bacteria to decompose. In addition to causing deoxygenation, some algal species produce toxins that contaminate drinking water supplies. Different treatment processes are required to remove nitrogen and phosphorus.

Nitrogen Removal

The removal of nitrogen is effected through the biological oxidation of nitrogen from ammonia (nitrification) to nitrate, followed by denitrification, the reduction of nitrate to nitrogen gas. Nitrogen gas is released to the atmosphere and thus removed from the water.

Phosphorus Removal

Phosphorus can be removed biologically in a process called enhanced biological phosphorus removal. In this process, specific bacteria, called polyphosphate accumulating organisms are selectively enriched and accumulate large quantities of phosphorus within their cells (up to 20% of their mass). When the biomass enriched in these bacteria is separated from the treated water, these biosolids have a high fertilizer value.

Disinfection

The purpose of disinfection in the treatment of wastewater is to substantially reduce the number of microorganisms in the water to make it potable for human use. The effectiveness of disinfection depends on the quality of the water being treated (e.g., cloudiness, pH, etc.), the type of disinfection being used, the disinfectant dosage (concentration and time), and other environmental variables. Cloudy water will be treated less successfully since solid matter can shield organisms, especially from ultraviolet light or if contact times are low. Common methods of disinfection include ozone, chlorine, or ultraviolet light.

Coastal Regulation Zone Notification, 1991.

In recent time, the coastal zones of world are under increasing pressure due to high rate of human population growth, development of various industries, mining, fishing, industrial waste effluents and discharge of municipal sewage. Such industrial development along the coast has resulted in degradation of coastal ecosystems and diminishing the coastal resources. Thus there is an urgent need to protect the coastal ecosystems and habitats by implementing the coastal regulation zone notification and integrated coastal zone management study. Healthy coastal life needs understanding and proper planning of environment, on and around the coasts. Perhaps due to this the Ministry of Environment and Forest, Government of India issued a notification in the year 1991, under Environment protection act of 1986, declaring coastal stretches as coastal regulation zone (CRZ) and thus regulating activities in CRZ.

India has a coast of about 7516 km long and 4198 islands are spread along the main coast of Andaman, Nicobar and Lakshadweep group. The coastal zone means the coastal water, wetland and shore land strongly influenced by marine water. This is the area of interaction between land and sea, which is influenced by both terrestrial and marine environment. The coastal zone includes the area between high tide line (HTL) and low tide line (LTL), up to 10 nautical miles towards the seaside from HTL and up to 20 km from HTL towards the land side. The accurate demarcation of shoreline is very important for planning purposes. The prime requisite of coastal regulation zone plan chart is basically to manage coastal and coastal zone features for sustainable use by demarcating high and low tide line on chart with the help of hydrographic surveys. The Ministry of Environment and Forests, Government of India issued the Coastal Regulation Zone Notification on 19th February 1991, under which coastal stretches were declared Coastal Regulation Zones (CRZ) and restrictions were imposed on the setting up and expansion of industries, operations and processes in the said Zones for its protection. The said notification has been amended from time to time based on recommendations of various committees, judicial pronouncements, representations from State Governments, Central Ministries, and the general public, etc., consistent with the basic objective of the said Notification.

Prohibited Activities:

The following activities are declared as prohibited within the Coastal Regulation Zone, namely:

i.   Setting up of new industries and expansion of existing industries, except

(a) Those directly related to water front or directly needing foreshore facilities and

(b) Projects of Department of Atomic Energy.

ii. Manufacture, handling, storage or disposal of hazardous substances as specified in the Notifications of the Government of India.

iii. Setting up and expansion of fish processing units including warehousing (excluding hatchery and natural fish drying in permitted areas); Provided that existing fish processing units for modernization purposes may utilize twenty five per cent additional plinth area required for additional equipment and pollution control measures.

iv. Setting up and expansion of units/mechanism for disposal of waste and effluents, except facilities required for discharging treated effluents into the water course with approval under the Water (Prevention and Control of Pollution) Act, 1974; and except for storm water drains.

v. Discharge of untreated wastes and effluents from industries, cities or towns and other human settlements.

vi. Dumping of city or town waste for the purposes of landfilling or otherwise.

vii. Dumping of ash or any wastes from thermal power stations.

viii. Land reclamation, bunding or disturbing the natural course of sea water except those required for construction or modernization or expansion of ports, harbours, jetties, slipways, bridges and sea-links and for other facilities that are essential for activities permissible under the notification or for control of coastal erosion and maintenance or clearing of water ways, channels and ports or for prevention of sandbars or for tidal regulators, storm water drains or for structures for prevention of salinity ingress and sweet water recharge: provided that reclamation for commercial purposes such as shopping and housing complexes, hotels and entertainment activities shall not be permissible;

ix. Mining of sands, rocks and other substrata materials, except

(a) Those rare minerals not available outside the CRZ areas and

(b) Exploration and extraction of Oil and Natural Gas

xi. Construction activities in ecologically sensitive areas as specified in this notification.

xii. Any construction activity between the Low Tide Line and High Tide Line except facilities for carrying treated effluents and waste water discharges into the sea, facilities for carrying sea water for cooling purposes, oil, gas and similar pipelines and facilities essential for activities permitted under this Notification; and

xiii. Dressing or altering of sand dunes, hills, natural features including landscape changes for beautification, recreational and other such purpose, except as permissible under this Notification.

Regulation of Permissible Activities:

1) Clearance shall be given for any activity within the Coastal Regulation Zone only if it requires waterfront and foreshore facilities.

2) The following activities will require environmental clearance from the Ministry of Environment and Forests, Government of India, namely:

- Construction activities related to projects of Department of Atomic Energy or Defence requirements

- Operational constructions for ports and harbours and light houses and constructions for activities such as jetties, slipways

- Thermal Power Plants

- All other activities with investment exceeding rupees five crores, except those activities which are to be regulated by the concerned authorities

(i) The Coastal States and Union Territory Administrations shall prepare, within a period of one year from the date of this Notification, Coastal Zone Management Plans identifying and classifying the CRZ areas within their respective territories.

7

Energy Resources

Energy the capacity of doing work is a primary input for almost all economic activities and is therefore vital for improving the quality of life. Its use in almost all the sectors has compelled us to focus our attention to ensure its continuous supply to meet our ever-increasing demands. The energy resources can be broadly divided into two categories.

1. Renewable sources of energy: these are the resources, which are available in unlimited amounts in nature and are continuously being produced in nature. These resources are inexhaustible means they can be renewed over a short period of time these resources include hydropower, solar power, wind power, tidal energy, geothermal energy, biomass energy and ocean energy etc.

2. Non-renewable sources of energy: these are the resources which are available in limited amounts and are developed over a longer period of time as million of years. They are finite and exhaustible means they cannot be renewed in short periods of time. These resources include fossil fuels (coal, petroleum and natural gas) nuclear fuels etc.

Limitations of conventional sources of energy: conventional sources of energy are those which are being most often used to meet out the energy demands. The fossil fuels are used more commonly than nonconventional sources of energy because most of the needs of energy of humans are being satisfied by burning of fossil fuels. But in some places the use of nuclear power and hydroelectricity has become dearer and are called as conventional sources at these places. Among all the conventional sources of energy coal occupied the central positions. During the last two decades the energy consumption has increased dramatically and these energy demands have been fulfilled by the so called conventional sources of energy are discussed below:

1. Exhaustibility: if the rate of their consumption increases more and more they are expected to exhaust (finish) in near future.

2. Pollution: their use besides exhaustibility is related to another problem called pollution. They add more and more undesirable and poisonous substances into the atmosphere leading to its disturbance along with the disturbance of humans and all other life.

3. Climate changes: the burning of fossil fuels releases many gases like CO_2, CO, SO_x and NO_x which upset the balance of our climate. Their excessive increase lead to the phenomena's like global warming, ozone depletion, acid rains and smog formation.

4. Accidents of the nuclear power plants at times lead to the emission of radioactive pollutants which are chronically hazardous to living organisms.

5. Related to them are a number of other environmental implications like:

   - Production of fly ash

- Thermal pollution of water bodies

- Safety and security risks

- Space heating etc.

Solar Energy:

It is the energy that originates from the sun by thermonuclear fusion reaction. The solar energy reaching earth is radiant energy, entering at the top of atmosphere at 1370w/m2-the solar constant. The energy ranges from UV to visible to IR (heat energy). About 50% of this energy reaches the surface of earth; 30%is reflected and the atmosphere absorbs 20%. The total amount of solar energy reaching earth is vast enough almost beyond belief. Although it is an abundant source, it is also diffuse (widely scattered) varying with seasons, latitude and atmospheric conditions. The solar energy received from the sun in the form of light radiations can be converted directly or indirectly into other forms of energy like heat and electricity, which can be utilized by humans through the thermal or the photovoltaic systems.

The solar thermal system indicated the use of the radiations of sun in the form of heat which in turn is converted into other forms of energy as electrical, mechanical and chemical etc. solar thermal devices include:

- Solar cockers

- Solar water heaters

- Solar dryers

- Solar stills

- Solar wood seasoning klins

- Solar air heaters etc.

The basic principle behind the solar thermal devices is the collection of the suns heat onto a blackened bottom of a box provided with a clear plastic top or glass top to prevent the heat escaping, as in a greenhouse. These devices are used for diverse purposes.

Photovoltaic system: it is the system in which solar energy is converted into electricity, thus providing an alternative to coal and nuclear power. Here, a few methods are feasible, but thee two important and economically viable methods are:

1) Photovoltaic cells: a solar cell, more properly called a photovoltaic cell of PV cell looks like a simple wafer of two very thin layers of material with one wire attached to the top and one to the bottom. The two layers of the cell are the semiconductor materials separated by a junction layer. The lower layer has atoms with single electrons in their outer orbits that are easily lost. The upper layer has atoms lacking single electrons in their outer orbits, these atoms readily give electrons. The kinetic energy of the striking photon on the sandwich dislodges electrons from the lower layer, creating an electric potential between the two layers. The potential provides the energy for an electric current to flow through the rest

of the circuit. Electrons from the lower side flow through a motor or some other electric devices back to the upper ide. Thus with no moving parts, solar cells convert solar energy directly into electric power with an efficiency of 20%. Since there are no moving parts in them, solar cells do not wear out.

Uses

i)   In pocket calculators, watches and toys

ii) Lighting homes as solar panels

iii) In irrigation pumps, traffic signals, radio transmitters, lighthouses and off-shore drilling platforms that are distant from power stations.

2) Solar trough collectors: these are the cost effective collector systems named so because the collectors are long trough shaped reflectors titled towards sun. the curvature of the trough is such that all the light hitting the collector is reflected onto a pipe running down the center of the system. Oil or some other heat absorbing liquid circulating through the pipe is thus heated to a very high temperature. The heated fluid is passed through a heat exchanger to boil water and produce steam for driving a turbo generator.

Besides these there are some more methods for the conversion of solar energy into electric energy like:

Power tower: a power tower is an array of sun tracking mirrors that focus the sunlight falling on several acres of land onto a receiver mounted onto a tower in the center of the area. The intense heat produces steam in the receiver (boiler), which drives a turbo generator and is expected to produce the electricity more economically than solar trough collector.

Advantages of Solar Energy:

i)   It is universal, decentralized and non-polluting

ii) It is available free of cost

iii) It is an unlimited source of energy

iv) It is bound to achieve greater economic importance in near future because of depletion of conventional resources

Limitations:

i)   It is an intermittent source of energy

ii) Its efficiency is variable depending upon the climatic conditions and varying weather phenomena's

iii) It is very expensive to build solar power stations

Wind Energy:

Moving air is known as WIND. It has energy. It is emerging as the most cost effective source of power because it combines the abundance of a natural source wind with modern technology. The interest in wind power increased throughout the world since the energy crisis of 1973-74. Wind is basically caused from the following two main features:

i)   Heating and cooling of atmosphere resulting in the convection currents.

ii) Rotation of earth with respect to the atmosphere and its motion around sun.

Since the energy possessed by wind is by virtue of its motion, so the device used to extract its energy should be capable of slowing down the wind. Wind energy conversion devices like wind turbines are used for converting wind energy into mechanical energy. A wind turbine consists basically of a few sails, vanes or blades rotating from a central axis. When wind blows against these blades they rotate about the axis. The rotational motion is then converted into some useful work. By connecting the wind turbine to an electric generator wind energy can be converted into electrical energy. The efficiency of the wind turbine depends upon:

i)   Wind speed (ideal speed is 4-25m/s)

ii) Cross section of the wind swept by the rotor

iii) Overall conversion efficiency of the rotor, transmission system and generator or pump

Wind-farm: an array of up to several thousand such turbines. The wind energy sector is growing fastest among all the renewable sources of energy throughout the world. In April 1999 worldwide wind energy capacity touched a record of 10000 megawatts. In India efforts to use wind as a form of energy began in 1985 with the establishment of wind mills in Gujarat with a capacity of 1.6MW. The Washington based world watch institute (WWI) recognized India as a wind super power. In India the potential of wind energy is available in Karnataka, Kerala, Tamil Nadu, Andhra Pradesh and the coastal areas of Gujarat and Maharashtra.

Uses of Wind Energy: it is used to

i)   Propel the sailboats in rivers and seas

ii) Run pumps to draw water from the ground

iii) Run flour mills to grind grains

iv) Generate electricity

v)   Scatter grains

vi) Separate grains from chaff

Advantages:

i)   It is available free of cost

ii) It is environment friendly and pollution free

iii) It is cheaper as there are no shortages of input costs

Limitations:

i)   It is intermittent, not available at all places and all times

ii) Wind mills are hazardous to birds

iii) They are extremely loud

iv) Large areas of land are required to produce useful amount of energy

Hydropower/Hydel Energy:

Hydel energy is the energy harnessed from the flowing and falling water. Water stored behind dams and at height having a huge amount of potential energy is released gradually and iss made to flow through channels under high pressure to drive hydraulic turbines and electric generators. The amount of power generated depends upon two things:

i)   The height of water behind the dam, that provides pressure to the water

ii) The volume of water that flows through the channels

Hydropower generates 17% of electric power throughout the world and is by far the most common form of renewable energy in use. In USA about 6.7% of the electricity comes from the hydroelectric dams.

Advantages:

i)   It is non-polluting source of energy and hence environment friendly

ii) It is a renewable source and thus saves the scarce fuel resources

iii) It is a reliable source with 90-95% availability

iv) The efficiency of hydro-generating units is much higher

v)   Hydrogenating power stations have longer life

vi) Cost of generation, operation and maintenance are much lower that other sources of energy

Although water power is having a number of characteristic features it still involves significant ecological, social and cultural tradeoffs:

i)   The reservoir created behind dams inevitably drowns the farmlands, wildlife habitats and perhaps towns or land of historical, archeological, geological or cultural importance

ii) They often displace the rural populations

iii) They impede or prevent the migration of fishes. Salmon fishing-one of the great industries of Northwest has been heavily affected

iv) They result in spread of parasitic diseases as the High Aswan dam of Egypt has fostered the spread of a parasitic worm that causes a debilitating disease

v) Ecological consequences occur below the dams

In India first hydropower generating unit was commissioned in 1987 in Darjeeling. We are having a great hydroelectric potential of 84000MW much of which is still untapped.

Tidal Energy:

The term tide is used for periodic rise and fall of waters of the oceans produced by the attraction of moon and sun. They result by the gravitational pull of sun and moon. These tides can be used to produce electric power called a tidal power. For this many imaginative schemes have been proposed but the most straight forward idea is the "tidal barrages" in which a dam is built across the mouth of a bay and turbines are mounted in the structures. The incoming tide flowing through the turbines would generate power. As the tide is shifted, the blades would be reversed so that the outflowing water would continue to generate power.

According to an estimate the total global tidal energy potential is around three million MW's of which India alone could generate 40000MW's from its coastline of Gujarat, West Bengal, Gulf of Cambay and Sunderbans. Other suitable sites are Andaman and Nicobar Islands and Lakshadweep islands.

Ocean Thermal Energy Conversion (OTEC):

Over much of the world oceans a thermal gradient of 20°C exists between the surface and deep-water zones. This temperature difference can be harnessed to produce power by an experimental technology known as Ocean Thermal Energy Conversion (OTEC). It involves the use of surface warm water to heat and vaporize a low boiling liquid such as ammonia. Thee increased pressure of the vaporized liquid would drive turbogenerators. The vapor leaving the turbines would then be recondensed by cold water pumped from as much as 100m deep and returned to start the cycle.

Studies have indicated that OTEC power plants show little economic promise unless perhaps they are coupled with other cost effective operations. Although it is nonpolluting and freely available but its low efficiency coupled with other drawbacks such as:

High capital costs,

Persistent maintenance,

Problems and fouling of pumps and pipelines rendered the OTEC technology uneconomical.

Biomass Energy:

Biomass is a term used for all materials originating from photosynthesis; it is basically the waste materials of the living organisms. It includes cattle dung, agricultural wastes, crop residues (bagasse or rice husk) and crops grown especially for energy content. Biomass resources can be converted into useful forms of energy and comes under these categories:

i) Solid traditional biomass: in this biomass is burnt directly to get heat energy

ii) Non-traditional forms of biomass: in this biomass is converted into methods and ethanol and used as liquid fuel

iii) Anaerobic fermentation: in this biomass is converted into gaseous forms like biogas by the process of fermentation.

The process of converting biomass resources into useful forms of energy is as follows:

i) Combustion: in this process the biomass with moisture content of about 15% is taken and direct combustion takes place. It can be used directly as heat as well as for the production of electricity.

ii) Anaerobic digestions: it is a process of allowing bacteria to feed on the biomass in absence of oxygen. The wastes are put into a biomass digester. In absence of oxygen a consortium of anaerobic bacteria breaks down the organic matter into methane, carbon dioxide, hydrogen, nitrogen and various other trace gases. The mixture of these gases is known as biogas or gobar gas (average compositions CH_4=55-57 %; CO_2=30-45 %; H_2S=71-2 %; N_2=1 % and others). The gross calorific value of this gas is about 5300 K Cal/m3. Because of its high methane content it is inflammable and is used as a source of fuel. The process takes place at the temperatures around upto 40°C and requires moisture content of about 80%. Besides biogas this process results in the formation of semisolid residues which is used as manure.

iii) Pyrolysis: it is the chemical decomposition of organic materials (biomass) by heating in absence of oxygen or any other reagent except possibly steam. By this the biomass is converted into a gas, fuelling the gas turbine for electricity generation. This is referred to as biomass gasification. plants.

iv) Petroplants: the plants which are the sources of liquid hydrocarbons-a substitute for liquid fuels are known as Petroplants. The hydrocarbons can be converted into petroleum hydrocarbons by specific processes. The plants belonging to Ephorbiaceae, Asclepiadaceae, Apocynaceae, Urticaceae, Convolvulaceae and Sapotaceae and over 385 species have been screened for hydrocarbon content. Some of the species of plants and trees proposed for energy plantation are Eucalyptus, Causurina, Sorghum, Accacia and Prosopis etc.

8

Environmental Impact Assessment (EIA)

Process of Environmental Impact Assessment (EIA):

The process of EIA should necessarily include:

1) A description of the proposed actions, it objectives along with the relevant technical information

2) The relationship of proposed action to the land use plans, policies and control for the affected areas

3) The probable impacts of the proposed action on the environment

4) Suggestions for alternate course of action to mitigate the undue impacts of the proposed action

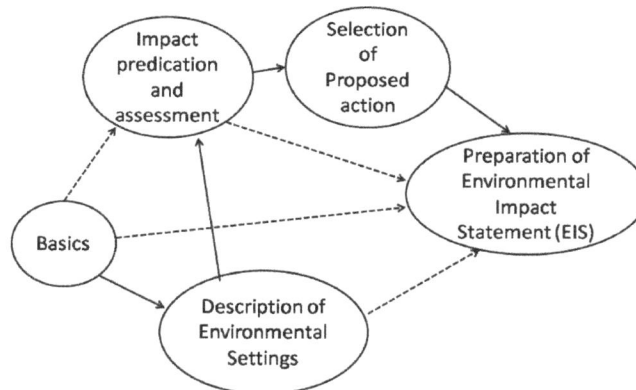

Process of Environmental Impact Assessment

Basics: It allows us to identify the need of the project and the general types of possible solutions. After the identification of the need for a project it is necessary to identify certain possible alternatives for meeting the project needs. It also includes the complete knowledge of the type of impacts associated with various alternatives. This step can be broken down into a few smaller steps as:

• Screening: whether an EIA is needed or not

• Scoping: identification of key issues and problems

• Identification and evaluation of alternatives: listing the alternative sites, techniques and the impact of each.

Description of Environmental settings: It involves the base-line information (of the area proposed for the activity) necessary to describe the predicted impacts associated with various alternatives under consideration. It is a list of environmental items developed on a project-by-project basis or at least a project type-by-type basis. One suggested arrangement for developing a list is to divide the environment into physical parameters, chemical parameters, biological parameters and socio-economic parameters or items.

Impact prediction and assessment: To predict and assess the impacts of a proposed action, it is necessary to describe the future of the environmental settings in the area without the project. Various techniques are available for projecting the future of these environmental settings. Prediction and assessment of the impacts of each alternative on physical, chemical, biological, cultural, and socio-economic environment are required. There are many scientific approaches and models that can be used to predict impacts on the air environment, water environment and noise environment. Most of the methods of impact analysis involve the concept of an impact scale and an impact importance. The predicted changes in air quality and water quality can be assessed against the environmental quality standards.

Following are the methods for impact prediction:

1) Mathematical or mechanistic models

2) Mass balance models

3) Statistical models

4) Physical, image or architectural models

5) Field and laboratory Experiment methods

Selection of proposed actions: Selection of the proposed action is made as per the guidelines of DOE (Department of Environment). First of all it involves the aggregation and display of the comparative environmental impacts of each alternative including the non-project alternative. Then the selection is done on the basis of:

- Professional judgment

- Public preferences (if public meetings are held)

- Cost benefit analysis

- Economic analysis

Preparation of Environmental Impact Statement (EIS):

The preparation of EIS is the last and important step of the EIA process. It involves:

- Preparation of draft EIS as per the guidelines of Department of Environment (DoE).

- Circulation of the draft EIS for review and comments to

    o Federal agencies

    o State and local agencies

      o   Public and private interest groups

- Incorporation of review comments

- Preparation of final EIS

- Submission of final EIS to relevant agency/decision making body

A filling period of 30-days is required prior to the initiation of the project action.

Methods of Impact Identification:

The impact methodology provides an organized approach for prediction an assessment of impacts. They provide an approach for evaluating absolute or relative impacts of alternatives. Following are some of the methods of impact analysis:

1) Adhoc method: It is the most common approach of impact assessment involving a team of experts or specialists to identify impacts in their areas of expertise. It is for the rough assessment of total impacts giving the broad areas of possible impacts and the general nature of these impacts.

2) Checklists: This is one of the basic methodologies used in EIA and is of various types:

   - Simple checklists: are the lists of parameters with no guidelines on how the data to be measures and interpreted.

   - Descriptive checklists: includes an identification of environmental parameters and guidelines about their measurement.

   - Scaling checklists: similar to descriptive checklists with the additional information basic to subjective scaling of parameter values.

   - Scaling weighting checklists: similar to scaling checklists with additional information about the subjective evaluation of one parameter value with respect to the other.

   - Questioner checklists: based on a set of questions to be answered. Some questions about the direct impacts and possible mitigation measures.

   - Threshold of concern checklists: consist of a list of environmental components, a threshold at which those assessing the proposal should become concerned with the impact.

3) Overlays: It is a well developed approach in planning and landscape architecture. It is based on the use of a series of overlay maps depicting environmental factors or land features e.g. physical, social, and aesthetic etc. A base map is prepared showing the general area within which the project may be located. Successive maps are then prepared for individual environmental components likely to be affected by the proposed action. By overlaying these maps the impacts are identified. The degree of impacts is shown by shading; with darker shading representing a greater impact and no shaded areas representing no impacts. Identification is done by superimposing the overlay maps.

4) Networks: They are based on the concept of known linkages within systems. They are an attempt to recognize that a series of impacts may be triggered by a project action. The actions associated with a project can be related to both direct and indirect impacts. They establish a cause-condition-effect relationship. The impact on one environmental factor may affect another environmental or socio economic factor and such interactions are listed on a network diagram. The diagram acts as a guide to impact identification and the presentation of results.

5) Matrices: They are basically generalized checklist where one dimension is a list of environmental, social and economic factors likely to be affected and the other dimension is a list of actions associated with the project. The matrices are of different types like:

   • Simple matrices: These are the two dimensional charts showing environmental items on one axis and developmental actions on other axis.

   • Magnitude matrices: These matrices describe the impacts of the project according to their magnitude and importance. They are also called as interaction matrices. The magnitude is the extensiveness and is described by assigning a numerical value from 1 to 10 with 10 representing a large magnitude and 1 representing small magnitude. Value near 5 represents an intermediate extensiveness. The importance is the significance of the assessment of the consequences. It is also given a numeric value ranging from 1 to 10 with 10 representing a very important interaction and 1 representing a very low interaction. One of the best known interaction matrices is the Leopold matrix developed by Leopold et al.

6) Quantitative or index method: This method is based on a list of factors thought to be relevant to a particular proposal and which are differentially weighed for the importance. Likely impacts are identified and assessed. Impacts are transformed into a common measurement unit for example a score on a scale of environmental quality.

Benefits of EIA:

1) It provides comprehensive coverage of short and long-term technical and social issues.

2) It generates alternatives, which are more acceptable to all parties involved.

3) It prevents adverse effects on environment.

4) It prevents technical troubles before start up by involving different parties in decision making.

5) It sets the basis for continuous monitoring of key projects and environmental conditions to prevent arising of any problem.

6) It adds new dimensions to existing planning policies.

EIA as an Effective Planning Tool:

Developmental projects whether small or large leave some adverse effects on the human environment. Activities such as construction, extraction of mineral and other natural resource, exploration of oil and gas, unsustainable agricultural practices etc affect the environment and most severely the human environment (encompassing the physical, chemical, biological and social). These can be grouped into two systems-the natural system consisting of aquatic, atmospheric, biological, geographical and geological compartments and the social system consisting of public health issues, environmental quality, cultural pluralism, environmental susceptibility, emotion, individual lifestyles, experiences, beliefs and values.

The EIA was developed as an effective planning tool for all these activities because it effectively protects, sustains and manages the environment (both physical and social) alongside the development and advancement. It works for the identification and evaluation of consequences of human actions on environment and when appropriate, mitigate those consequences or at least minimizes or compensates the consequences.

Social Impact Assessment (SIA):

Social impact assessment (SIA) is understood in different ways and consequently many writers do not give a specific definition of SIA. But obviously it has something to do with understanding the impacts of a project or policy on people. It can also be defined, as an effort to assess or estimate, in advance the social consequences that are likely to follow from specific government actions (including buildings, large scale projects and leasing large tracts of land for resource extraction).

It can also be defined as the prediction and assessment of impacts of the policies and actions in context with the social setup of the individuals of a society i.e. the changes in the norms, values and beliefs of the people. Predicting the changes in one or more of the following assesses the social impacts:

- People's way of life: how they live, work, play and interact with one another on a day-to-day basis.

- Their culture: beliefs, customs and values

- Their community: its cohesion, stability, character and facilities

- Their environment:

    o   Quality of air and water the people use

    o   Level of dust and noise they are exposed to

    o   Adequacy of sanitation

    o   Safety and fears about their security

    o   Access to and control over resources

So, in general SIA is concerned with everything that affects people, their lives, their culture, their mental and physical health and their community and that represents a change that may have been caused either directly or indirectly, by a specific project or a policy as well as with the cumulative impacts of several projects. Benefits of SIA: There are three primary benefits to use SIA:

1) It is a system which ensures both equity and transparency of decision making

2) It is a form of risk assessment; in that the identification of likely impacts of development can be assessed in calculations related to a project to ensure that the future costs of mitigation do not exceed the benefits of the projects.

3) Through the participation process it can lead to better decision making by incorporating local knowledge which in turn leads to reduced impacts, both in terms of physical impacts as well as perceived impacts, stress, uncertainty and notions of fairness.

Some other Benefits of SIA are:

- It provides for transparent decision-making
- It is a potential safeguards against corruption
- By incorporating public participation it allows better site planning
- It is a democracy at work

Limitations of SIA: Burdge and Vauclay (1995) identified four major problems affecting SIA:

1) There are difficulties in applying social sciences to SIA because it does not have an applied tradition. 2) There are difficulties with the SIA process itself particularly in terms of lack of effective techniques, tools and models and the quality of available data.

3) There are problems with the procedures applying to SIA particularly because administrative structures have not facilitated the use of SIA. There are no formal review processes and evaluation practices.

4) There exists a bias amongst various groups of people that lead to be a perception by developers, regulatory agencies and sometimes by environmental professionals, that social issues are unimportant. This leads to lack of commitment to SIA.

Process of Social Impact Assessment (SIA):

The process of SIA is completed in a number of steps and some major steps are:

1) Public involvement: It means to develop and implement an effective public involvement plan to involve all the potentially affected groups of people like those:

   o   Who live nearby

   o   Who are forced to relocate

   o   Who have interest in the new project

   o   Who are the owners of the land on which the project is proposed

2) Identification of Alternatives: It means to describe the proposed action and reasonable alternatives. At a minimum detail will be required about:

- o   Locations

- o   Land requirements

- o   Need for ancillary facilities (Road, water supply, electric supply)

- o   Construction schedule

- o   Work schedule

- o   Work force (during both constructional and operational phases)

3) Baseline conditions: It means a document for the relevant human environment relations. The social baseline conditions prior to effects due to the current project. This includes an understanding of:

- o   The relationship between social and biophysical environment

- o   The historical background of the area

- o   Political and social structure of the area

- o   Cultural back ground

- o   Attitude and psychological conditions of people

- o   Basic population characteristics

4) Scooping: After obtaining the technical understanding of the project, the identification of the full range of possible social impacts through a variety of means including discussions or interviews with members of all potentially affected people.

5) Projection of estimated impacts: It involves the evaluation of all possible impacts to determine the probable impacts.

6) Predictions of responses to impacts: it involves the determination of the importance of the identified social impacts to the affected public.

7) Estimation of indirect and cumulative impacts: it considers the flow on ramification of projects including the second order impacts. Also considers how the impacts of one project may affect and be affected by the other projects.

8) Changes in alternative: it recommends new or changed alternatives and estimates the consequences.

9) Mitigation: it means to develop mitigation plans. Mitigation plan should:

- o   Firstly seek to avoid impacts.

- o   Secondly seek to minimize unavoidable impacts.

- o   Thirdly utilize compensation mechanism.

10) Monitoring: It means to develop a monitoring plan that is capable of identifying deviations from the proposed action and any important unanticipated impacts. It should also be able to compare the real impacts with the projected ones.

9

Satellite Imaging as a Mean of Monitoring the Global Environment

Satellite Imaging:

It is the employment of satellite to obtain information in the form of images about any process, phenomenon or area. Remote sensing means to get the information about anything without coming in contact with that particular thing. It can be defined in a number of ways as:

I. It is the Art and Science of obtaining information about an object without coming in contact with that object.

II. Remote Sensing is the technique of acquiring data about an object without touching it physically.

III. Remote Sensing is a tool similar to mathematics using the sophisticated sensors to measure the EMR's emitted by an object from a distance and then interpreting the information by using mathematically and statistically based logarithms.

IV. Remote Sensing is defined as the science of collecting and interpreting information about a target without being in physical contact with the target under study.

Remote sensing is of various categories starting from the observations by naked eyes, photography by cameras, and photography from aircrafts and sensing the sensors from space satellites. The information obtained as EMR's by means of instruments like cameras, sensors, scanners, lasers and aircrafts is recorded and is then analyzed by means of visual and digital image processing e.g. when we read a book our eyes act as sensors which respond to the light reflected from the book and hence we are able to interpret the information present in the book.

Types of Remote Sensing:

Depending upon the source of energy, which illuminates the object under study, the remote sensing technique has been classified into two types:

i) Active Remote Sensing

ii) Passive Remote Sensing

i) Active Remote Sensing: It is the type of remote sensing in which the naturally reradiated or reflected energy from the earth's surface features is measured by the sensors operating in different selected spectral bands. It is similar to photography in daytime without flash.

ii) Passive Remote Sensing: It is that remote sensing technique in which the remote sensing system supplies its own source of energy to illuminate the object and measures the reflected energy returned to the system. It is similar to photography during nighttime with flash.

Application of Remote Sensing:

Remote sensing data is non-selective with respect to information content i.e. the data interpreted from individual record may be applicable to diverse fields. The information is useful to investigators in many diverse disciplines as mentioned below.

1. Geologists used remote sensing to find deposits of minerals and petroleum.

2. Soil scientists list out the important characteristics of soil by relating them to geological and geographical features.

3. Foresters and agriculturalists determine the kind of trees and plants growing in an area, their health condition and yield.

4. Hydrologists locate useful aquifers and estimate the volume of surface flow in the watersheds.

5. Geographers analyze land use pattern, climate, topography, plant life, animal life and human activities in a particular area.

6. Wild life managers use it for locating habitats of various kinds of animals; the photographs often show violation of the law. e.g. illegal mining in remote areas, pollution of water by dumping of chemical and effluents from poisonous gases through chimneys of industries.

7. The disasters like floods, hurricanes and fires can be assessed by the study of remote sensing data and the information so obtained can be used in making emergency decisions and in combating the situation.

8. Fast development in industrialization throughout the world is causing serious problems of environmental pollution. Employing the remote sensing technique can effectively do monitoring of this. The remote sensing has multidisciplinary applications in monitoring the global environmental conditions.

Hence for the effective management of diminishing natural resources and monitoring of environment, the remote sensing has got an edge over the conventional methods.

Necessity and Importance of Remote Sensing:

The dramatic increases in the population and the rising living standards have been mounting a tremendous pressure on the natural resources resulting in their loss. It therefore becomes necessary to manage the available resources effectively. For the effective management of the natural resources periodically inventories were made on ground involving different people like geologists surveying for minerals, foresters and agriculturists looking for trees and crops in the forests and fields, surveyors walking the countryside in the course of preparing the necessary maps. But the data obtained by them was inaccurate, time consuming and costly. So, the advent of satellite imagery was a big step forward as the data obtained from this was more accurate, less time consuming and less expensive. In this technique surveying/ sensing is done simultaneously in several bands of the electromagnetic spectrum. In this way much more information about an area can be secured as compared to that obtained with conventional photography, which is limited to visible band of spectrum.

Remote Sensing-Appraisal and Management of Natural Resources:

It is an established fact that planning for integrated development depends upon the reliable and comprehensive appraisal and integration of natural resources. The appraisal of natural resources such as land forms, soils, vegetation and water resources through conventional methods is unreliable, time consuming and costly. Remote sensing techniques through visual interpretation of aerial photographs, satellite imagery and digital analysis of computer compatible tapes (CCT's) in conjunction with ground truth have been faster, reliable and economic for the appraisal and integration of natural resources. Although LANDSAT imagery in the form of black and white and color transparencies and computer compatible tapes have been found very useful for this purpose. The IRS satellite images due to their higher resolution ranging from 50.8 to 72.5m have much better potentialities.

Landforms act as an environment for the development and distribution of soil, growth of vegetation and circulation of surface and ground water. For the classification and mapping of landforms the remote sensing techniques have been found economical, faster and accurate. Landforms being most conspicuous are distinctly visible on aerial photographs and satellite imagery. Environmental hazards like wind erosion, water erosion, salinity/alkalinity and water logging associated with different land forms could also be easily identified and mapped for systematic and rational environmental planning. The drainage channels and drainage pattern can also be identified and mapped by remotely sensed data.

10

Technology and Environment

Technology:

It is a term with its origin in Greek "technologia" with "techne" meaning craft and "logia" meaning saying. It is a broad concept that deals with a species usage and knowledge of tools and crafts and how it affects a species ability to control and adapt to its environment. Technology can be most broadly defined as the entities both material and immaterial created by the application of mental and physical efforts in order to achieve some value. It can also be used to refer to a collection of techniques and knowledge of how to combine resources to produce desired products, to solve problems, fulfill need s or satisfy wants, it includes technical methods, skills, processes, techniques, tools and raw materials. It is the technical means people use for fulfilling their needs.

Technological innovation and evolution increases the mastery of man over resources. It has affected societies and their surroundings in a number of ways, in many societies it has helped develop more advanced economies and has allowed the rise of leisure class. Many technological processes produce unwanted by-products, known as pollution and deplete natural resources to the detriment of earth and its environment. Technology causes social changes and can be viewed as an activity that forms or changes culture.

Throughout the 20th century, the use of the term has increased to the viewpoint where it now encompasses different types of technologies:

Technology as Objects: tools, machines, and instruments, weapons appliances that are the physical devices of technical performance.

Technology as knowledge: it includes the technical know-how behind the technological innovations.

Technology as Activities: it includes the activities of the people their skills, methods, procedures and routines.

Technology as Process: it begins with a need and ends with a solution.

Technology as Socio-technical system: it includes the manufacture and use of objects, which involves people and other objects in combination.

Milestones of Technological Evolution:

1. Use of tools: dating back to 2.5 million years ago. This era of stone tools is called "Paleolithic" or "old stone age"

2. Discovery, utilization and control of fire: the exact date of this discovery is not known.

3. Discovery of agriculture: it allowed for feeding of large populations and a transition to a sedentary lifestyle.

4. Application of steam engine: refined in 1700's, they were increasingly fuelled by coal. Meanwhile it found wide industrial applications by converting heat energy into mechanical energy. It is considered to be the most important single innovations for industrialization.

5. Control of Nuclear power: it was achieved a few decades ago and is used to generate electricity in societies where coal is scarce. Although it is efficient, the fuel produced is expensive and hazardous and is having unprecedented potential for destruction and thus imposing a risk to our survival.

Nature of technology: technology has a number of characteristic features, which are as follows:

1. Relation with science: often the two terms are confused. Science deals with the natural world and technology deals with the human made world. Science deals with the understanding while as technology deals with the doings. Science is a reasoned investigation of phenomena and technology is often a consequence of science and engineering.

2. It is multidimensional: it involves co-operation between different specialists e.g. production engineers and material scientists. It many involve technologists also in performing a multitude of functions.

3. It is socially shaped: technological enterprises are determined neither by advances in knowledge nor by the identification of needs, but by social interests. It is shaped by consumer choice of the society.

4. It is concerned with values: technology is informed by values at every pint. Value decisions are being involved for the designs and solutions being called right or wrong in ethical terms.

Classification of technology: the technology has been classified into different types. Some of the types of technology are as follows:

1. Low-tech

2. High-tech

3. Traditional technology

4. Alternate technology

5. Intermediate technology

6. Appropriate technology

7. Blended technology

8. Hard technology

9. Soft technology

1. Low-tech: these are the technologies which are:

    •   Small scale

    •   Simple in infrastructure

    •   Simple to use

    •   Easy to operate

    •   Less costly to construct and obtain

    •   Mostly labor intensive and do not involve skills to handle.

2. High-tech: these are the technologies which are:

    •   Large scale

    •   Complex in infrastructure (sophisticated)

    •   Difficult to use and operate

    •   Often costly to construct and obtain

    •   Not labor intensive, requiring very little but skilled human inputs

    •   Generally semi-automatic to automatic in nature

    •   Based on machines

3. Traditional technology: they are indigenous technologies, which encompass traditional techniques accumulated over generations based on local values, rituals, beliefs and knowledge. The term indigenous technology has been defined as that which originates, grows or lives naturally in a particular place. It is known as locally generated/ home grown technology, which is born naturally in a land, pertaining to or intended for the natives. It thus means the technology that is developed within a society and is not brought out from outside and adopted.

Traditional technologies form the fabric of culture. They are those that have evolved with a community and have been passed from one generation to another e.g. techniques of managing fields, cultivation, milking, milk storage and fermentation techniques.

4. Alternate technology: the technology which represents the viable second choice to existing mainstream technology is called as alternate technology e.g.

    •   Small scale organic farming instead of chemically intensive cultivation

    •   Renewable energy in place of non-renewable energy

5. Intermediate technology: technology that lies mid-way between traditional technology and modern technology is called as intermediate technology. It is significantly more effective and expensive than traditional methods but cheaper than developed world technologies. It is a technology that the proponents argue can be easily purchased and used by poor people

and can lead to greater productivity e.g.

- Ox drawn plough in-between the traditional hoe and modern tractor

- Kerosene stove between the traditional chulla and the LPG stove

6. Appropriate technology: it is a technology that is designed with special considerations to the environmental, ethical, cultural, social and economic aspects of community it is intended for. Appropriate technologies are congenial with their surroundings and are characterized by organizational simplicity, high adaptability to a particular environment, sparing the natural resources and low costly final products. They are the best fit in any particular scenario and have high potential for employment e.g. biogas digesters.

7. Blended technology: appropriate technologies designed for culturally sensitive introduction in societies other than that of which it originates is called as blended technology. It is adapted to the norms and values of local cultural conditions.

8. Hard technology: the tools, implements, machines, devices and equipment's that are the physical embodiment of technology are called hard technologies. The hard technology includes engineering techniques; physical structures and machinery that meets a need defined by a community and utilize the material readily available.

9. Soft technology: the social structures, human interactive processes, motivation techniques, design methodologies, support system and decision making processes which form the arrangement for individual and collective self-determination are called as soft technologies.

Appropriateness of Technology:

The appropriateness of technology is judged by the effects of the technology on the well being of individuals, on environment and on communities. When designing, it is important to critically consider the appropriateness of their proposed products in terms of their positive and negative impacts on people and environment. For judging the appropriateness of a technology the following factors are taken into consideration:

1) Aesthetic appropriateness: it refers to a products sensory appeal for a user. It involves the form, function, shape, color, material, space, texture and design of the final products of a technology.

2) Cultural appropriateness: it refers to the need and requirements of different groups and sub-groups with a society based on their beliefs, practices and heritage.

3) Functional appropriateness: it involves that how well the products of a technology match with their users. It also includes a consideration of the efficiency and effectiveness of a product.

4) Social appropriateness: it refers to the manner in which products are shaped as per the needs of the societies and groups within those societies.

5) Economic appropriateness: it involves the economic viability of a product by taking into

consideration its cost and benefits.

6) Environmental appropriateness: it involves the balance between benefits of a new product and its impacts on the environment.

7) Ethical appropriateness: it refers to the possible effects of technology in relation to the views and acts of individuals and societies.

Environmentally Sound Technologies:

Environmentally sound technologies (EST's) are those technologies, which have the potential for significantly improved environmental performance relative to other technologies. Broadly speaking the technologies:

1. Protect the environment

2. Are less polluting

3. Use resources in a sustainable manner

4. Recycle more of their wastes and products

5. Handle all residual wastes in a more environmentally acceptable way than the technologies for which they are substituted.

6. Generate low or no wastes

7. Cover end-of-pipe technologies as well

8. Are compatible with the society and environment in which they are to be used

9. Are relatively simple and understandable in infrastructure

10. Are suitable to local maintenance and repair.

Furthermore environmentally sound technologies are not just individual technologies, but total systems which include know-how, procedures, goods and services, equipment's as well as organizational and managerial procedures i.e. they are the full spectrum of production and consumption technologies that are more ecofriendly that the technologies for which they are the substitutes.

- The adoption of EST's involves the application of:

- Ecological principles

- Cleaner production: a recognized and proven strategy for

    a) Improving the use of natural resources

    b) Reducing and eliminating the wastes at source

    c) Minimizing the potential risks to human health

- Monitoring and assessment systems for environment

- Pollution prevention and control technologies

- Ecological engineering to strike a balance between engineering principles and environmental considerations

Interaction of Technology, Society, Environment, Development and Society

Technology is a term more often used to describe the knowledge and achievement in the area of practical outcome. It contributes to the changes in cultural, social, environmental and economic circumstances. The changes are positive (developmental) if the technology is appropriate as well as indigenous because the cross-cultural sharing of technology leads to unexpected results. The appropriateness of technology is judged by its effect on the well being of individuals, environments and communities. The indigenous technology comes from within a culture and is more suitable for it than a foreign innovation. The interaction of technology and culture is cyclic, as it arises and becomes a part of the culture, changes it, and creates a new void in it, which is to be fulfilled by a new technology. It changes the environments and then we need a new technology to get adjusted in the changed environments.

The technology influences and shapes the society. It leads to progressions in the society. Some technologies bring fortunes like Information Technology, Biotechnology and Nanotechnology etc. to man while some others bring misfortunes like the nuclear weapons. Actually it is not the technology itself but the inappropriate use of the technology, which is responsible for the problems and had let deep scars on the society.

Technology is having a close proximity with development as well. Without technology no economic development, no social development, no political development and no progress is possible because technology is synonymous to productivity, development, progress and efficiency.

Technology has greatly influenced the environmental sector as well by leading to the increased incidences of:

- Global warming (by the excessive emission of GHG's by fossil fuel burning in automobiles and industries)

- Ozone layer depletion (by the innovation and invention of CFC's as the better and safer alternatives to sulpher and ammonia)

- Lung cancer with asbestos fibers (by making it as an alternative fire proofing agent in schools, offices and other buildings)

- Generation of different types of hazardous wastes

- Acid rain formation (by the excessive emission of sulpher oxides and nitrogen oxides into the atmosphere)

Besides this certain technological innovations, which are inappropriate, have led to the destruction of the natural resource base. So to undo the deadly impacts of inappropriate technologies the need of the hour is to:

1. Select the appropriate technologies

2. Reorient the technologies as per the needs of the people

3. Control our technologies

4. Maintain the technology transfer as per the society's potentials

5. Blend the technologies into appropriate places

6. Implement the technology only after knowing its scientific background and to conduct a lot of research over it

7. Choose the technology not only on economic considerations and sophistication but on the basis of its cultural, economic, environmental, energy and social standards

8. Interchange the technology properly i.e. technology should not be readily interchanged between first world (urban) and third world (rural) countries, because if not well understood creates problems.

Need to Reorient Technology:

Technological development is having a direct role in the evolution of environmental problems like:

- Production of pollutants from both the production sector as well as the consumer sector

- Depletion of ozone layer

- ardous wastes

- Solid wastes and their proper management

So the urgent need of the hour is to reorient the technology so that it becomes appropriate and sustainable. By reorienting the technology we can:

- Reduce the extent of environmental problems

- Bring our industries to zero emission levels by preventing pollution and designing for nature

- Make the production and consumption sectors ecofriendly

- Go for waste minimization and pollution prevention

Steps for Reorientation of Technology:

1. Development of appropriate and sustainable technology (AST's) i.e. technologies which are:

    - Small scale

    - Energy efficient

- Environmentally sound

- Labor intensive

- Simple to maintain

- Matching the user

2. Throw away concept of hazardous wastes has to be given out

3. Environmental elegance should be an integral part of engineering education and practice so as to design environmentally compatible technologies.

4. Environmental quality should be incorporated into the selection and design of manufacturing processes

5. Environmental engineering should become an integral part of all branches of engineering

6. Bringing engineering education, practice, industrial priorities and public policy into alignment to eliminate the paradox of technological development

7. Emphasis should be laid down on "Front end control" rather than on "End of pipe control"

8. Emphasis should be given to waste minimization by:

- In-plant recycling

- Substitution of inputs

- Changing the process

- Modification of the end products

Biotechnology:

It is defines as a technique that uses living organisms to make or modify a product, to improve plants and animals or to develop microorganisms for specific uses (OR) a technology using biological phenomena's for copying or manufacturing various kinds of useful products (OR) the controlled use of biological agents such as microorganisms or cellular components for beneficial uses (OR) the application of biological systems, organisms and processes to manufacturing and service industries.

Actually biotechnology is having its roots in molecular biology and microbiology. It encompasses a wide array of specialized disciplines right from the age-old fermentation processes to the latest techniques of genetic engineering. It has received tremendous attention in recent years due to its immense potentialities and application in diverse fields like aquaculture, agriculture, immunology, forestry, chemical production and pollution control etc. Genetic engineering is the fundamental basis for the modern biotechnology. Genetic engineering involves the ability to transfer specific genes from one kind of organism to other thereby producing new biological properties.

Biotechnology has got diverse application and potentials. Some of them are as follows:

1. Biotechnology and Pollution Control: biotechnology comprises of integrated application of theoretical and practical aspects of biochemistry, microbiology, physiology, genetics and chemical engineering to exploit the properties of microbes and cell cultures for various beneficial technological uses. Environmental biotechnology is a very broad field which includes environmental monitoring and safety, waste treatment and recovery, restoration of environmental quality, substitution of non-renewable resource base with renewable resources, research and development of various process for the benefit of mankind with due regard to socioeconomics.

2. Biosensors: A biosensor is an analytical gadget comprising of an immobilized layer of a biological material (e.g. enzymes, antibiotics, hormones, nucleic acids or whole cells) in conjunction with a transducer which analyses the biological signals and converts them into an electrical signal. The layer of suitable biological material is immobilized on a permeable membrane, which is kept in the close vicinity of a sensor. The substance to be measured passes through the membrane and interact with the immobilized material and yield a product. The product passes through another membrane to the transducer. The transducer converts the product into an electric signal, which is amplified and displayed or recorded.

3. Treatment of toxic wastes using genetically improved organisms (GMO's) Dr. Anand Mohan Chakerborthy successfully synthesized an oil eating superbug by introducing plasmids from four different strains as OCT, XYL, CAM and NAH plasmid by successive conjugation into a single cell of "Pseudomonas putida". This superbug can degrade all the four types of substrates which are as follows:

   OCT plasmid degrades oxane, hexane and decane

   XYL plasmid degrades xylenes and toluenes

   CAM plasmid degrades camphor

   NAH plasmid degrades naphthalene

   These otherwise would have required all the four separate plasmids. In 1990, these microbes were used for cleaning up of an oil spill in water in Texas (USA).

4. Recovery of useful products from waste materials e.g. recovery of CH_4, metals etc. From waste materials.

5. Development of new catalysts, new bioreactors, novel biosensors and automation of wastewater treatment processes.

6 Biotechnology and crops: modern biotechnology has revolutionized the crops by the development of transgenic, speeding up of the plant growth.

7. What takes place on a small scale under controlled laboratory conditions is completely different from what occurs on a large-scale real world situation.

8. The genetically altered foods have the possibility of causing allergic reactions in a low percentage of people.

9. The potential environmental impacts of transgenic crops could be negative such as impacts of Bt. Crops (i.e. genetically modified crops producing toxins of the soil bacteria Bacillus thuringiansis) causing increased resistance to Bt. Toxin by the pest. Crop losses due to killing non-target bio-control organisms, reduction in soil fertility due to Bt. Toxins remaining in the soil and potential transfer of insecticidal properties or virus resistance to wild relatives of crops are some of the damages that we may face.

Legal Regimes for Sustainable Development National Legislative Framework for Environment

Environmental concern in India is as old as the Indian civilization. Traditionally the environmental concerns here were mostly aimed at conservation of forests, wildlife and other natural resources only. But after independence the issues of main concern shifted to sanitation, public health and hygienic disposal of community wastes as these issues resulted in many outbreaks of deadly communicable diseases. However, now the concern has broadened to include sustainable development. The period from independence to 1972 was a period of continuous deterioration and degradation due to two reasons:

i) Large scale continuity of activities like industrialization, urbanization, rapid population growth, poverty, intensification of agriculture, deforestation, loss of wildlife, and pollution of air, pollution of water and soil.

ii) No serious concern was given to these issues, but the only concern was given to public health and sanitation.

Issue of environmental protection was addressed for the first time in the 4th five year plan (1969-1974). The year 1972 is very significant in the history of environmental conservation because this year 24th General Assembly of United Nations decided to convene an international conference on Environment and Development and a national committee on Environmental planning and coordination was also formed. The conference was held in Stockholm and was attended by the then prime minister of India Smt. Indra Gandhi. By her participation in the said conference there was a paradigm shift in the concept and concern for environmental protection and conservation. Innumerable steps were taken at every level of social, cultural, political, administrative, governmental and non-governmental organization for the protection of environment.

Environmental Legislation:

Environment legislations consist of all the legal guidelines that are intended to protect our environment. The primary objective of these legislations is to prevent the environment from any kind of damage. Here the objective of environmental protection is achieved by a strict code of conduct-framed with a blend of preventive, promotional and mitigative measures. In general there are three types of legislations relating to environment viz. Planning legislation, protective legislation and preventive legislation. But in most cases legislations are created for protective and preventive purposes. However, planning legislation for environmental quality improvement is almost lacking.

Environmental Legislation in India:

Environmental legislation is not a new idea for India but it was present here even in the pre-independence period. Bengal Smoke Nuisance Act-1905 is perhaps one of the oldest environment protection acts in India followed by the Factories Act 1948. Similarly Mines and Minerals (Regulation & Development) Act 1957 is another important environmental legislation formulated in post-independence period. Even before India's independence in 1947, several environmental legislations existed but the real impetus for bringing about a well-developed framework came only after the United Nations Conference on Human Environment (Stockholm, 1972). A series of acts were passed by the parliament thereafter. Under the influence of this declaration, the National Council for Environmental Policy and Planning within the Department of Science and Technology was set up in 1972. This Council later evolved into a full-fledged Ministry of Environment and Forests (MoEF) in 1985 which today is the apex administrative body in the country for regulating and ensuring environmental protection. After the Stockholm Conference, in 1976 constitutional sanction was given to environmental concerns through the 42nd Amendment, which incorporated them into the Directive Principles of State Policy and Fundamental Rights and Duties. Since 1970's an extensive network of environmental legislation has grown in the country. The MoEF and the pollution control boards (CPCB i.e. Central Pollution Control Board and SPCBs i.e. State Pollution Control Boards) together form the regulatory and administrative core of the sector.

A policy framework has also been developed to complement the legislative provisions. The Policy Statement for Abatement of Pollution and the National Conservation Strategy and Policy Statement on Environment and Development were brought out by the MoEF in 1992, to develop and promote initiatives for the protection and improvement of the environment. The EAP (Environmental Action Programme) was formulated in 1993 with the objective of improving environmental services and integrating environmental considerations in to developmental programmes. Other measures have also been taken by the government to protect and preserve the environment. Several sector-specific policies have evolved in India as well. India has also adopted a federal system in which power of the Govt. is shared between the centre and state. Part XI of the constitution from articles 245-263 regulates the legislative and administrative relations between centre and state regarding the issues of environment.

- Article 245-This article empowers parliament to make laws for the whole or any part of the territory of India, and state legislatures to make laws for their respective states or any part of the State.

- Article 246-This article divides various areas of environment and legislation between the union and state. It divides them into three lists-the union list, the state list and the concurrent list. It also provides that if a conflict arrives on any item between state and centre, the central law shall prevail.

  i) Union list: It comprises of 97 subjects over which only centre has exclusive powers of legislation. It includes industries, mineral resources, atomic energy etc.

  ii) State list: It comprises of 66 subjects over which states have exclusive powers of legislation subject to their territorial limitation. It includes the areas like public health, sanitation, hospitals, water supply, drainage and fisheries.

iii) Concurrent list: It comprises of 47 subjects over which both parliament and state legislatures have powers of legislation. The areas present in this list include forests, population control, family planning, and protection of wild animals and birds etc.

Thus the distribution of legislative powers between Center and State under federal system of government provides enough provisions to make laws dealing with various environmental problems. The legal regimes for environmental protection available in India are broadly comprises of:

- Common law Remedies and

- Stationery Remedies.

Common Law Remedies:

Common law is the law that exists independently of any legislation. It is a set of laws and principles that has been continually developed by courts over hundreds of years. It is important to realize that "common law" is not a fixed or absolute set of written rules in the same sense as statutory or legislatively enacted law. There are no specific common law actions designed to protect the environment, as the common law has principally developed to protect the individual's rights and private property rights. However, when an environmental impact also interferes with an individual's right or a private property right, the common law can be used to protect the environment indirectly. For this reason, generally only a person whose interests have actually been affected by the harm can bring an action under the common law. Common law remedies against environmental pollution are available under law of torts. It is among the oldest legal remedies to abate pollution. Tort is basically a civil wrong other than a breach of trust. A breach of the common law is said to give rise to a "cause of action". Some common law causes of action that might be used to protect the environment are:

- Trespass;

- Private nuisance;

- Public nuisance; and

- Negligence.

Each of these causes of action can be used to protect different rights in different situations, although it is often the case that several causes of action may be applicable to a single issue. The common law may be used to protect the environment instead of statute because it may provide a legal recourse that is not otherwise available under the legislation, or because in some situations it may provide remedies that are more desirable or suitable than those available under legislation.

Trespass

Trespass occurs where a person directly, intentionally or negligently and without permission causes some physical interference with another person's property. For example, if pesticides being used on your neighbour's property accidentally blow onto your land, it is unlikely that this would constitute trespass. The trespass may be committed by casting material upon another's land, by discharging water, soot or carbon, by allowing gas or oil to flow underground into someone else's land, but not by mere vibrations or light which are generally classed as nuisances.

In the case of *Martin vs. Reynolds Metal Co* the deposit on Martin's property of microscopic fluoride compounds, which were emitted in vapor form from the Reynolds' plant, was held to be an invasion of this property—and so a trespass.

Nuisance:

Anything that annoys hurts or interfaces with the quality of life is referred to as nuisance. A person of his own property, working an obstruction or injury to the right of another or to the public, and producing such material annoyance, inconvenience, and discomfort that the law will presume resulting damage defines nuisance." A nuisance arises whenever a person uses his property to cause material injury or annoyance to a reasonable neighbor. Odors, dust, smoke, other airborne pollutants, water pollutants and hazardous substances have all been held to be nuisances. Nuisance actions come in two forms: public and private. Private nuisance is committed where one person ("the defendant") substantially and unreasonably interferes with another person ("the plaintiff")'s right to the use and enjoyment of their land. "Public nuisance" occurs when a person causes a nuisance which "endangers the life, health, property morals or comfort of the public" or "obstructs the public in the exercise or enjoyment of rights common to all Her Majesty's Subjects".

Negligence:

When there is a duty to take care of and it has not been taken into consideration which results in harm to another person it is called negligence. "Negligence" is "the omission to do something which a reasonable man, guided by those ordinary considerations which ordinarily regulate human affairs, would do, or the doing of something which a reasonable and prudent man would not do. Negligence is that part of the law of torts which deals with acts not intended to inflict injury. Negligence may also be used as a cause of action to address environmental harm.

Nissan Motor Corp. vs. Maryland Shipbuilding and Drydock Company exemplifies a negligence action in an environmental case. The shipbuilding company's employees failed to follow company regulations when painting ships, allowing spray paint to be carried by the wind onto Nissan's cars. The shipbuilders had knowledge of the likely danger of spray painting, yet failed to exercise due care in conducting the painting operations in question. This failure to exercise due care amounted to negligence.

Strict Liability:

If a person for his/her own purpose bring to his/her own land, collets and keeps what which is likely to be harmful to another, he/she is to compensate for the damage done. This is helpful in hazardous waste based industries in India. It ensures that the burden of damage is to be beard by the industry concerned. After Bhopal Gas tragedy it was replaced by "Absolute liability" articulated by Supreme Court and adopted by parliament.

Statutory Law Remedies:

These are the laws of national and regional legislatures and local governments. Various activities which are harmful to environment can be controlled through this Act. They include:

Indian penal code, 1860: It is a code of conduct of Indian constitution to eradicate different types of wrong from the society. Its chapter XIV comprises of Sections 268-294A in which:

i)   Indian penal code, 1860: It is a constitution to eradicate different types of wrongXIV comprises of Sections 268-294A in which: code of conduct of Indian s from the society. Its chapter

Chapter XIV : offences affecting the public health, safety, convenience, decency and morals	
268.	Public nuisance
277.	Fouling water of public spring or reservoir
278.	Making atmosphere noxious to health
290.	Punishment for public nuisance

Any violation shall be a punishable with an imprisonment up to 3 months and a fine up to Rs 500 or both.

ii) Criminal procedure code, 1973: It is another strict code of conduct which can be involved to prevent any type of pollution. Its chapter X Part B contains Sections 133-143 and Part C has the directions for speedy and effective remedy against pollution problems and nuisance under section 133 CrPC. If a person fails to comply with its provision, he/she is prosecuted under section 183 of IPC. It can also be used against all statutory bodies like municipal corporation and govt. bodies etc. Thus both IPC and CrPC, although of ancient vintage, are contemporary remedial weapons.

Constitutional Perspectives:

Protection of environment prior to 42nd amendment was availed through section 21 of the constitution-stating that "no person shall be deprived of his life or personal liability except according to procedures established by law" i.e. ensuring that every persons has fundamental right to life that comprises right to live in a healthy environment, clean air.

Fundamental Duties:

The constitutions 42nd Amendments Act, 1976 added a new Part IV A dealing with "fundamental duties" in the constitution of India. Article 51A (g) specifically deals with the fundamental duties with respect to the environment and provides:

"It shall be the duty of every citizen of India to protect and improve the natural environment including forests, rivers and wildlife and to have compassion for living creatures"

The fundamental duties are intended to promote peoples participation in restructuring and building a welfare society. The protection of environment is a matter of constitutional priority. Environmental problem is the concern of every citizen and neglect of it is an invitation of disaster. Article

51A (g) gives the right to the citizens to move to court to see that the state performs its duties faithfully in accordance with the law of the land.

The true scope of Article 51A (g) has been best explained by Rajasthan high court in L.K. Koolwal vs. State of Rajasthan in 1988, a case where in the municipal authority was charged with the "Primary duty" to clean public streets, place and sewers and all spaces, not being private property which are open to the enjoyment of public, removing of noxious vegetation and all public nuisance and to remove filth, rubbish, night soil, odour or any other noxious or offensive matter. But it failed to perform its "Primary duty" resulting in acute sanitation problems in Jaipur posing a threat to human life. Mr. Koolwal moved to High court on the basis of the right given to it by Article 51A (g) for the enforcement of the duty cast on state.

Directive Principles of State Policy:

Part IV of the constitution deals with the directive principles of state policy. The constitutions (42nd Amendment) Act added a new directive principle in Article 48A dealing with the improvement of environment. It provides:

"The state shall endeavor to protect and improve the environment and to safeguard the forests and wildlife of the country"

The state cannot treat the obligation of protecting and improving the environment as mere pious obligation. The directive principles are not mere show pieces in the window dressing. They are *"fundamental in the governance of the country"* and they being part of the supreme law of land have to be implemented in letter and spirit. This incorporation of putting obligations on the *"state" as well as "citizens"* to protect and improve the environment in the *"Suprema Lax"* is certainly a positive development of Indian law. In *Shri Sachindanand pandey vs. State of West Bengal* the supreme court pointed out that whenever a problem of ecology is brought before the court, the court is bound to bear in mind article 48A and 51A(g) of the constitution.

State Laws:

State has enacted the laws on various issues of the environment like water pollution, and air pollution but in accordance with the central laws in those particular areas to tackle various facets of the pollution problems. The state laws are

The Water (Prevention and Control of Pollution) Act, 1974

Water is a state subject and as such Central Govt. cannot make laws relating to it, provided the state legislature authorizes it to do so under clause (I) of Article 252 of the constitution. After seeking the consent of 12 states, parliament passed a legislation regarding water called as water (prevention and control of pollution) Act. *"An Act to provide for prevention and control of water pollution and the maintaining or restoring of wholesomeness of water, for the establishment, with a view to carrying out the purposes aforesaid, of Boards for the prevention and control of water pollution, for conferring on and assigning to such Boards powers and functions relating thereto and for matters connected therewith"*

This Act represented India's first attempts to comprehensively deal with environmental issues.

The Act prohibits the discharge of pollutants into water bodies beyond a given standard, and lays down penalties for non-compliance. The Act was amended in 1988 to conform closely to the provisions of the EPA, 1986. It set up the CPCB (Central Pollution Control Board) which lays down standards for the prevention and control of water pollution. At the State level, the SPCBs (State Pollution Control Boards) function as per the direction of CPCB and the state government.

Objectives of the Act:

1. Prevention and control of water pollution.

2. Sustaining or restoring the wholesomeness of water.

3. Establishment of Boards for prevention and control of water pollution.

Functions of Central Pollution Control Board: The functions of the Central Pollution Control Board as given by Section 16-A of the act are

i)   To promote cleanliness of streams and wells in different area

ii) To advise the central govt. on various issues of pollution

iii) To co-ordinate the work of state Boards and to resolve the disputes among them

iv) To provide technical assistance and support to SPCB

v)   To establish laboratories for water analysis

vi) To lay down standards for steams and wells

Functions of State Pollution Control Boards: The functions of the State Pollution Control Boards as given by Section 7B of the act are

i)   Planning a comprehensive programme for prevention, control abatement of water pollution.

ii) To advise state govt. On various issues of pollution

iii) To collaborate with CPCB for handling the issues of pollution

iv) Evolving economical and reliable methods for disport and treatment of waste water

v)   To establish and recognize laboratories of water analysis

vi) To lay down standards for various activities

vii) Conducting research work for prevention and control of water pollution.

Powers of the State Boards:

i)   Power to obtain information

ii) Power to take samples

iii) Power to obtain a report of the result of analysis by a recognized laboratory

iv) Power of entry and inspection for performing the entire duty

v) Power of prohibition on disposal of polluting matter into a stream or well

Besides, the act provides for a permit system, or consent procedure to prevent and control water pollution. It prohibits the disposal of wastes into the water bodies in excess of the standards established by the boards. It also provides that if a person is interested to establish any industry he/she has to get consent from the state and if a person fails to do so the board may order closure, prohibition or regulation of any industry. The boards are also empowered to stop or regulate electric supply, water supply and other facilities to the unit.

Penalties:

Whoever fails to comply with any direction given under sub-section (2) or sub-section (3) of section 20 within such' time as may be specified in the direction shall , on conviction, be punishable with imprisonment for a term which may extend to three months or with fine which may extend to ten thousand rupees or with both and in case the failure continues, with an additional fine which may extend to five thousands rupees for every day during which such failure continues after the conviction for the first such failure. Whoever:

(a) Destroys, pulls down, removes, injures or defaces any pillar, post or stake fixed in the ground or any notice or other matter put up, inscribed or placed, by or under the authority of the Board.

(b) Obstructs any person acting under the orders or directions of the Board from exercising his powers and performing his functions under this Act,

(c) Damages any woks or property belonging to the Board,

(d) Fails to furnish to any officer or other employee of the Board any information required by him for the purpose of this Act,

(e) Fails to intimate the occurrence of an accident or other unforeseen act or event under section 31 to the Board and other authorities or agencies as required by that section,

(f) In giving any information which he is required to give under this Act, knowingly or willfully makes a statement which is false in any material particular,

(g) For the purpose of obtaining any consent under section 25 or section 26, knowingly or willfully makes a statement which is false in any material particular, shall be punishable with imprisonments for a term which may extend to three months or with fine which may extend to 1 [ten thousand rupees] or with both.

The Air (Prevention and Control of Pollution) Act, 1981:

This act was passed under article 253 of Indian constitution in pursuance of the decisions of Stockholm conference. To counter the problems associated with air pollution, ambient air quality standards were established, under this Act. The Act provides means for the control and abatement of air pollution. It seeks to combat air pollution by prohibiting the use of polluting fuels and substances, as well as by regulating appliances that give rise to air pollution. Under the Act establishing or operating of any industrial plant in the pollution control area requires consent from state boards. The boards are also expected to test the air in air pollution control areas, inspect pollution control equipment, and manufacturing processes. Objectives of the Act: The objectives of this act are

1) To provide for the prevention, control and abatement of air pollution in order to preserve the quality of air.

2) To provide the powers to the boards to take a deterrent action on the accused and performing certain functions necessary to improve environment.

The Central Board for the Prevention and Control of Water Pollution central Board constituted under section 3 of the Water (prevention and control of pollution) Act, 1974 (6 of 1974), shall, without prejudice to the exercise and performance of its powers and functions under that Act exercise the powers and perform the functions of the Central Board for the Prevention and Control of Air Pollution under this Act.

Powers and Functions of Boards:

Subject to the provisions of this Act, and without prejudice to the performance, of its functions under the water Prevention and Control of Pollution) Act, 1974 the main functions of the Central Board shall be to improve the quality of air and to prevent, Control or abate air pollution in the country.

The Central Board May:

(a) Advise the Central Government on any matter concerning the improvement of the quality of air and the prevention, control or abatement of air pollution.

(b) Plan and cause to be executed a nation-wide programme for the prevention, control or abatement of air pollution;

(c) Co-ordinate the activity of the States and resolve disputes among them;

(d) Provide technical assistance and guidance to the State Boards, carry out and sponsor investigations and research relating to problems of air pollution and prevention, control or abatement of air pollution;

(e) Plan and organize the training of persons engaged or to be engaged in programmes for the prevention, control or abatement of air pollution on such terms and conditions as the Central Board may specify;

f) Organize through mass media a comprehensive programme regarding the prevention, control or abatement of air pollution;

g) Collect, compile and publish technical and statistical data relating to air pollution and the measures devised for its effective prevention, control or abatement and prepare manuals, codes or guides relating to prevention, control or abatement of air pollution.

h) Lay down standards for the quality of air,

i) Collect and disseminate information in respect of matters relating to air pollution;

(j) The Central Board may establish or recognize a laboratory or laboratories to enable the Central Board to perform its functions under this section efficiently.

(k) The Central Board may delegate any of its functions under this Act generally or specially to any of the committees appointed by it.

The Functions of a State Board Shall be:

(a) To plan a comprehensive programme for the prevention, control or abatement of air pollution and to secure the execution thereof;

(b) To advise the State Government on any matter concerning the prevention, control or abatement of air pollution;

(c) To collect and disseminate information relating to air pollution:

(d) To collaborate with the Central Board in organizing the training of persons engaged or to be engaged in programmes relating to pfevention, control or abatement of air pollution and to organize mass-education programme relating thereto;

(e) To inspect, at all reasonable times, any control equipment, industrial plant or manufacturing process and to give, by order, such directions to such persons as it may consider necessary to take steps for the prevention, control or abatement of air pollution;

(f) To inspect air pollution control areas at such intervals as it may think necessary, assess the quality of air therein and-take step for the prevention, control or abatement of air pollution in such areas;

(g) To lay down, in consultation with the Central Board and' having regard to the standards for the quality of air laid down by the Central Board standards for emission of air pollutants into the atmosphere from industrial plants and automobiles or for the discharge of any air pollutant into the " atmosphere from any other source whatsoever.

(h) To advise the State Government with respect to the suitability of any premises or location for carrying on any industry which is likely to cause air pollution;

(i) To perform such other functions as may be prescribed or as may, from time to time, be entrusted to it by the Central Board or the State Government;

(I) To do such other things and to perform such other acts as it may think necessary for the proper discharge of its functions and generally for the purpose of carrying into effect the purposes of this Act.

(2) A State Board may establish or recognize a laboratory or laboratories to enable the State Board to perform its functions under this section efficiently.

Penalties and Procedure

whoever fails to comply with the provisions of section 21 or section 22 or directions issued under section 31 A, shall, in respect of each such failure, be punishable with imprisonment for a terms which shall not less than one year and six months but which may extent to six years and with fine and in case the failure continues, with an additional fine which may extend to five thousand rupees for every day during which failure continues after the conviction for the first-such failure.

If the failure referred to in sub-section (1) continues beyond a period of one year after the date of conviction, the offender shall be punishable with imprisonment for a term which shall not be less than two years but which may extend to seven years and with fine.

Whoever:

(a) destroys, pulls down, removes, injures or defaces any pillar, post or stake fixed in the ground or any notice or other matter put up, inscribed or placed, by or under the authority of the Board,

(b) obstructs any person acting under the orders or directions of the Board from exercising his powers and performing his functions under this Act,

(c) damages any works or propane belonging to the Board,

(d) fails to furnish to the Board ,or any officer or other employee of the Board if any information required by the Board or such officer or other employee for the purpose of' this Act,

(e) fails to intimate the occurrence of the emission of air pollutants into the atmosphere in excess of the standards laid down by the State Board or the apprehension of such occurrence, to the State Board and other prescribed authorities or agencies as required under sub-section (1) of section 23,

(f) in giving any information which he is required to give under this Act, makes a statement which is false in any material particular,

(g) for the purpose of obtaining any consent under section 21, makes a statement which is false in any material particular shall be punishable with imprisonment for a term which may extend to three months or with fine which may extend to ten thousand rupees or with both.

National laws: The national laws present in the constitution of India for the protection and improvement of environment are

The Environment Protection Act, 1986:

This Act (may be called the environment (protection) Act 1986), came into force on 19 Nov. 1986 in the whole Indian under Article 253 of Indian constitution in pursuance with the declarations of Stockholm conference. It is an umbrella legislation to provide powers and framework for central govt. to coordinate activities of various central and state authorities established under previous laws. It was promulgated to provide for the protection and improvement of environment and the matters connected there with. Under this Act, the central government is empowered to take measures necessary to protect and improve the quality of the environment by setting standards for emissions and discharges; regulating the location of industries; management of hazardous wastes, and protection of public health and welfare.

Objectives: The objectives of the environment (protection) Act 1986 are:

a) Protection and improvement of environment

b) Implementation of the decisions of united nations conference on human environment

c) Covering the uncovered gap in the areas of environment

d) Coordination of the work of various agencies

e) Providing a deterrent punishment to those who endanger the safety of environment and health

f) Maintenance of harmony between environment and human beings.

The Act empowers the central govt. to take all necessary measures for the protection and improvement of environment like:

• Laying down standards for environmental safety

• Laying down standard for emission or discharge of environmental Pollutants

• Laying down procedures and safeguards for prevention of article

• Laying down procedures for handling hazardous wastes

• Designation of certain of nationwide programmers for preventing, controlling & obtaining pollution

• Establishment of environmental laborites

• Constitution of authority to exercise its powers

• Appointment of officers, their powers and functions

• Preparation of manuals, codes, guides related to pollution

Power of central Govt. Subject to different provisions of this Act, the Central Government shall have power to take all such measures as it deems necessary or expedient for the purpose of protecting and improving the quality of the environment and preventing, controlling and abating environmental pollution.

In particular and without prejudice to the generality of provision of subsection (1) such measures may include measures with respect to all or any of the matters namely:

1) Co-ordination of actions by the State Government, officer and other authorities.

2) Planning and extension of nation-wide programme for the prevention, control and abatement of environmental pollution.

3) Laying down standards for the quality of environment in its various aspects;

4) Laying down standards for emission or discharge of environmental pollutants from various sources whatsoever:

5) Restrictions of areas in which any industries, operations or process, or class of industries, operations or processes shall not be carried out or shall be carried out subject to certain safeguard.

6) Laying down procedures, safeguards for the prevention of accidents, which may cause environmental pollution and remedial measures for such accidents.

7) Laying down procedures and safeguards for the handling of hazardous substances:

8) Examination of such manufacturing processes, material and substance as are likely to cause environmental pollution.

9) Carrying out and sponsoring investigations and research relating to problems of environmental pollution.

10) Inspection of any premises, plants, equipment, machinery, manufacturing or other processes, materials or substances and giving by order, of such direction to such authorities, officers and persons as it may consider necessary to take steps, for prevention, control and abatement of environmental pollution.

11) Establishment or recognition of environmental laboratories and institutes to carry out functions entrusted to such environmental laboratories and institutes under this Act;

12) Collection and dissemination of information in respect of matters relating to environmental pollution;

13) Preparation of manual, codes or guide relating to the prevention control and abatement of environmental pollution;

14) Such other matters as the Central Government deems necessary or expedient for the purpose of securing the effective implementation of the provision of this Act

15) Powers of entry and inspection – for performing any function entrusted under legislation

16) Power to take samples – of air, water, soil from the factory for analysis.

17) Power to establish laboratories for analysis purposes

18) Power to give direction:

- The closure, prohibition or regulation of any industry, operation or process; or

- For stoppage or regulation of the supply of electricity or water or any other service.

Penalties:

Any person failing in any of the provisions of this Act shall be punishable with imprisonment for a term which may extend up to 5 years or with fine which may extend up to rupees one lakh or both. In case the violation or contravention continues beyond a period of one year after the date of conviction, the offender shall be punishable with imprisonment for a term of seven years.

Rules notified under environment (protection) Act: variousrules notified under EPA-1986 are:

Biomedical waste (Management and Handling) Rules, 1998:

In exercise of the powers conferred by section 6, 8, 25 of the EPA, 1986 the central government notified the rules for the management and handling of bio-medical wastes. This was published on 20th July 1998 and appeared in the official gazette of India on 27th July 1998. These rules apply to all persons who generate, collect, receive, store, transport, dispose or handle the bio-medical wastes in any form. These rules regulate the disposal of all types of bio-medical wastes including blood, body parts, medicines, glass, solid wastes, animal wastes, liquids and biotechnological wastes. The Biomedical waste means any waste, which is generated during the diagnosis, treatment or immunization of human beings or animals or in research activities pertaining thereto or in the production or testing of biological and including categories mentioned in schedule I of the Rules.

Objective: The main objective of these rules is to take steps to ensure safety to health and environment.

Duty of the Occupier: It shall be the duty of every occupier of an institution generating bio-medical wastes which includes a hospital, nursing home, clinic, dispensary, veterinary, animal house, pathological laboratory, block bank by whatever name called, to take all steps to ensure that such waste is handled without any adverse effect to human health and environment.

Treatment and Disposal

1) Bio-medical waste shall be treated and disposed of in accordance with Schedule I, and in compliance with the standards prescribed in Schedule V.

2) Every occupier, where required, shall set up in accordance with the time-schedule in Schedule VI, requisite bio-medical waste treatment facilities like incinerator, autoclave, microwave system for the treatment of waste, or, ensure requisite treatment of waste at a common waste treatment facility or any other waste treatment facility.

Segregation, Packaging, Transportation and Storage:

1) Bio-medical waste shall not be mixed with other wastes.

2) Bio-medical waste shall be segregated into containers/bags at the point of generation in accordance with Schedule II prior to its storage, transportation, treatment and disposal. The containers shall be labeled according to Schedule III.

3) If a container is transported from the premises where biomedical waste is generated to any waste treatment facility outside the premises, the container shall, apart from the label prescribed in Schedule III, also carry information prescribed in Schedule IV.

4) Notwithstanding anything contained in the Motor Vehicles Act, 1988, or rules there under, untreated biomedical waste shall be transported only in such vehicle as may be authorized

for the purpose by the competent authority as specified by the government.

5) No untreated bio-medical waste shall be kept stored beyond a period of 48 hours The pollution control boards and pollution control committees are the prescribed authorities for the implementation of these rules.

Municipal Solid Waste (Management and Handling) Rules, 2000:

In exercise of the powers conferred by the EPA, 1986 the ministry of environment and forests. Government of India notified the municipal solid wastes (Management and handling) Rules 2000. These rules were notified on 25th September 2000, with the aim to take all the necessary steps to properly manage and handle the municipal solid wastes so as to protect the human health and environment. These rules make the municipal bodies/local bodies responsible for the management and handling of these wastes. It includes four main schedules.

Schedule I: In this deadline have been given for:

i) Setting up waste processing and disposal facilities

ii) Monitoring the performance of these facilities

iii) Improving existing and identification of new landfill sites for future use.

Schedule II: describes the standards for

i) Segmentation

ii) Collection

iii) Storage

iv) Transformation

v) Processing

vi) Disposal of solid wastes

Schedule III: it provides specifications for

i) Site selection

ii) Development and establishment of sanitary landfills

iii) Landfill closure and post case once it is covered iv) Controlling air, water pollution and monitoring standards

Schedule IV: Indicate waste processing options including; standards for composting, treated leachates and incinerations.

Recycled Plastic Manufacturing and Usage Rules, 1999:

In exercise of the powers conferred by the EPA–1986, The ministry of environment and Forests, Government of India notified the plastic manufacturing and usage rules 1999, for regulating the manufacturing and use of recycled plastic carry bags and containers.

Features of the Rules:

1) The usage of carry bags and containers made of recycled plastic, for food item is strictly prohibited.

2) For the manufacture of carry bags and containers, made of plastic, the following conditions should be satisfied.

   i) Carry bags and containers made of virgin antacids should be in natural shade or white.

   ii) Carry bags and containers made of recycled plastic shall be manufactured using pigment and colorants as per IS 9833: 1981, entitled *"list of pigments and colorants used in plastics in contact with foodstuffs, pharmaceuticals and drinking water"*

   iii) Recycling shall be undertaken according to Bureau of Indian standards specification: IS 14534: 1998, entitled: *"The guidelines for Recycling of plastics"*

   iv) Marking and codification of carry bags and containers shall also be as per BIS specification: IS 14534: 1998, entitles *"The Guidelines for recycling of plastics"*

   v) Manufacturers shall print on each packet of carry bags as to whether these are made of "recycled material" or *"virgin material"*

   vi) Minimum thickness of carry bags shall not be less than 20 microns.

   vii) The plastic industry Association, through their member units, shall undertake self regulatory measures.

Prescribed Authority:

a) The prescribed authority for enforcement of the provisions of these rules related to manufacture and recycling shall be the State Pollution Control Boards in respect of States and the Pollution Control Committees in respect of Union Territories;

b) The prescribed authority for enforcement of the provisions of these rules related to the use, collection, segregation, transportation and disposal shall be the District Collector/Deputy Commissioner of the concerned district where no Such Authority has been constituted by the State Government/Union Territory administration under any law regarding non-bio-degradable garbage.

Noise Pollution (Regulation and Control) Rules, 2000:

Whereas the increasing ambient noise levels in public places from various sources, like industrial activity, construction, fire crackers, sound producing instruments, generator sets, loud speakers, public address systems, music systems, vehicular horns and other mechanical devices have deleterious effects on human health and the psychological well being of the people; it is considered necessary to regulate and control noise producing and generating sources with the objective of maintaining the ambient air quality standards in respect of noise. In exercise of the powers conferred by the EPA, 1986, the central Govt. introduced one more set of rules namely the noise pollution (Regulation and control) Rules in 2000 to control noise pollution.

Objective: the main objectives of the set of rules are:

i)   Regulation and control of noise producing and generating sources.

ii) Maintaining ambient air quality standards with respect to noise.

Features: Various features of the Noise Pollution (Regulation and control) Rules include

1) The State govt. shall categorize the areas into silence, residential, commercial and industrial zones/areas to implement noise standards for them

2) The State Government shall take measures for abatement of noise including noise emanating from vehicular movements, blowing of horns, bursting of sound emitting firecrackers, use of loud speakers or public address system and sound producing instruments and ensure that the existing noise levels do not exceed the ambient air quality standards specified under these rules.

3) An area comprising not less than 100 metres around hospitals, educational institutions and courts may be declared as silence area / zone for the purpose of these rules.

4) The noise levels in any area / zone shall not exceed the ambient air quality standards in respect of noise as specified in the Schedule.

5) The authority shall be responsible for the enforcement of noise pollution control measures and the due compliance of the ambient air quality standards in respect of noise.

6) The respective State Pollution Control Boards or Pollution Control Committees in consultation with the Central Pollution Control Board shall collect, compile and publish technical and statistical data relating to noise pollution and measures devised for its effective prevention, control and abatement.

7) A loud speaker or a public address system or any sound producing instrument or a musical instrument or a sound amplifier shall not be used at night time except in closed premises for communication within, like auditoria, conference rooms, community halls, banquet halls or during a public emergency.

8) The noise level at the boundary of the public place, where loudspeaker or public address system or any other noise source is being used shall not exceed 10 dB (A) above the ambient noise standards for the area or 75 dB (A) whichever is lower;

9) The peripheral noise level of a privately owned sound system or a sound producing instrument shall not, at the boundary of the private place, exceed by more than 5 dB (A) the ambient noise standards specified for the area in which it is used.

10) No horn shall be used in silence zones or during night time in residential areas except during a public emergency.

11) Sound emitting fire crackers shall not be burst in silence zone or during night time.

12) Sound emitting construction equipments shall not be used or operated during night time in residential areas and silence zones.

Consequences of any Violation in Silence Zone / Area:

Whoever, in any place covered under the silence zone / area commits any of the following offence, he shall be liable for penalty under the provisions of the Act:

1) whoever, plays any music or uses any sound amplifiers,

2) whoever, beats a drum or tom-tom or blows a horn either musical or pressure, or trumpet or beats or sounds any instrument, or

3) Whoever, exhibits any mimetic, musical or other performances of a nature to attract crowds.

4) whoever, bursts sound emitting fire crackers; or

5) Whoever, uses a loud speaker or a public address system.

Complaints to be Made to the Authority:

1) A person may, if the noise level exceeds the ambient noise standards by 10 dB or more given in the corresponding columns against any area / zone or, if there is a violation of any provision of these rules regarding restrictions imposed during night time, make a complaint to the authority.

2) The authority shall act on the complaint and take action against the violator in accordance with the provisions of these rules and any other law in force.

Ambient Air Quality Standards in respect of Noise

Area Code	Category of Area / Zone	Limits in dB(A) Leq*	
		Day Time	Night Time
(A)	Industrial area	75	70
(B)	Commercial area	65	55
(C)	Residential area	55	45
(D)	Silence Zone	50	40

Note:- 1. Day time shall mean from 6.00 a.m. to 10.00 p.m.
2. Night time shall mean from 10.00 p.m. to 6.00 a.m.
3. Silence zone is an area comprising not less than 100 metres around hospitals, educational institutions, courts, religious places or any other area which is declared as such by the competent authority
4. Mixed categories of areas may be declared as one of the four above mentioned categories by the competent authority.

The Wildlife (Protection) Act, 1972, Amendment 1991

The Wildlife Protection Act, 1972, provides for protection to listed species of flora and fauna and establishes a network of ecologically-important protected areas. The WPA empowers the central and state governments to declare any area as wildlife sanctuary, national park or closed area. There is a blanket ban on carrying out any industrial activity inside these protected areas. It provides for authorities to administer and implement the Act; regulate the hunting of wild animals; protect specified plants, sanctuaries, national parks and closed areas; restrict trade or commerce in wild animals or animal articles; and miscellaneous matters. The Act prohibits hunting of animals except with permission of authorized officer when an animal has become dangerous to human life or property or so disabled or diseased as to be beyond recovery. The near-total prohibition on hunting was made more effective by the Amendment Act of 1991.

The Forest (Conservation) Act, 1980

This Act was adopted to protect and conserve forests. The Act restricts the powers of the state in respect of de-reservation of forests and use of forestland for non-forest purposes (the term non-forest purpose includes clearing any forestland for cultivation of cash crops, plantation crops, horticulture or any purpose other than re-afforestation).

Convention on Climate Changes:

Formally the discussion of convention on climatic changes began at the international level in 1988 by the establishment of Intergovernmental Panel on Climate Changes (IPCC) by United Nations Environmental Programme (UNEP) and World Meteorological Organization (WMO). It is a panel consisting of thousands of scientists from hundreds of countries established to assess the current stage of knowledge about climate changes. The IPCC's first assessment report played an important role in adoption of the United Nations Framework Convention on Climate Changes (UNFCCC) at Rio-Earth summit in 1992. The convention agreed to the goal of stabilizing the greenhouse gas level in atmosphere, starting by reducing green house gas emission to 1990 level by the year 2000 in all industrialized nations. As the treaty was lacking specific emission targets, the countries were to achieve the goals by voluntary means. The convention made it clear that the developed countries must take lead in combating the climate changes and the adverse effects thereof. The parties must take precautionary measures, should cooperate and promote supportive open economic system leading to sustainable development.

Obligations on Developed Countries:

   i)   To return to 1990 level of GHG's by the year 2000

   ii) To prepare a national communication with specific information on policies

   iii) To promote, facilitate and finance the transfer of EST's (environmentally sound technologies) to developing countries

Obligation on Developing Countries:

   i)   To prepare an inventory of GHG's

   ii) To prepare a general list of the steps to implement the convention

   iii) To undertake sustainable development as per the convention provisions

The convention also designed GEF (Global Environment Facility) as an interim mechanism for the financial assistance to support its implementation in developing countries. The convention also established institutional mechanism for periodic review and an update of commitments, including the scheduling of regular conferences. Promoted by coalition of island nations, the third conference of parties to the UNFCCC met in Kyoto Japan in Dec. 1997 to craft a binding agreement on reducing green house gas emissions. In this protocol (Kyoto protocol) the parties agreed to reduce emission of six green house gases to 5.2% below 1990 levels, to be achieved by 2012.

Convention on Bio-diversity/Biological Diversity (CBD):

Although CITES provides some protection to some species of flora and fauna but, it is inadequate to address broader issues pertaining to biological diversity. After several years of recognition, the Convention on Biological Diversity (CBD) was drafted and became one of the pillars of the 1992 Earth summit in Rio-de Jenerio. The Biodiversity treaty as the convention is called was ratified in December 1993 and is now in force. The treaty established a COP (Conference of Parties) as the agency that will provide oversight and report on its task during its periodic meetings.

Objectives of the Convention:

i)   Conservation of biological diversity

ii) Ssustainable use of the components of biodiversity

iii) Fair and equitable sharing of benefits arising out of the utilization of genetic resources

The Signatory Countries, Must:

- Adopt specific national biodiversity action plans and strategies

- Establish a system of protected areas and ecosystems within their respective countries

- Establish policies that provide incentives to promote sustainable use of biological resources

- Restore habitats that have been degraded

- Protect threatened species

- Respect and preserve the knowledge and practices of indigenous peoples

- Respect the ownership of genetic resources by countries and share the technologies developed from those resources

- Promote the awareness about the importance of biological diversity through media and educational programmes

- Protect and conserve biodiversity as it stays within the sovereignty of the nations

United Nations Convention on the Law of Seas:

There are certain areas of land, water and air that do not fall in the national boundaries called as country's exclusive economic zones. The United Nation Convention on the Law of Seas (UNCLOS) held on December 10, 1982 established certain duties regarding the marine environments. The main features of UNCLOS are as follows:

i) The member countries of the convention have strict obligations to protect and preserve the marine environments and take necessary measures for pollution control.

ii) The member countries should regulate the activities such that they are not harmful to the seas.

iii) The nations should cooperate at global and regional basis with international organizations to formulate rules, standards, guidelines and procedures to protect marine environment.

iv) It promotes promotion of scientific research and data exchange preferences on marine environments.

v) It calls upon the nations to establish appropriate scientific criteria for the formulation of international environmental rules, standards, practices, procedures and safeguards for prevention, reduction and control of pollution caused to seas.

Space Treaty:

The outer space treaty signed in 1967, commands the parties to pursue studies of outer space including the moon and other heavenly (celestial) bodies. It permits conducting exploration of these heavenly bodies, so as to avoid harmful contamination and adverse changes in earth's environment due to introduction of extraterrestrial matter and adopt necessary measures for this purpose.

Convention on Forests:

The issue of an international forestry convention with an international binding character, raised by European Economic Commission (EEC) led by Germany and supported by USA, was very much controversial. The EEC and USA pursued a global forest treaty which would have been a binding treaty. Developing countries strongly opposed and challenged the binding treaty for that would have meant compromising with their sovereign rights to exploit their own resources. They argued that the developed world was trying to globalize the natural resources of the developing countries while allowing its own forest to be ruthlessly exploited on the plea that these were privately owned. India and other developing countries strongly opposed the move and took the stand that our forests could no longer be available as grounds for dumping hazardous wastes by the developed would. The convention was finally signed by various countries reflecting the first global convention on forests. It was also decided to keep the forests under regular assessment for International Corporation.

The United Nations Conference on Environment and Development (UNCED) also known as "Earth Summit" took place in June 1992, in Rio de Janerio, the capital of Brazil. It marks the 20th anniversary of the first United Nations conference on Human Environment (UNCHE) which took place between June 5-10, 1992 in Stockholm Sweden. The conference was attended by the leaders and representatives from 180 countries. Sustainable development was the primary focus of this summit. The UNCED was guided by the remarkable document of 1987 i.e. the Brundtland report and ended with the adoption of certain crucial documents-a "blueprint" intended to guide development in sustainable direction into and through the 21st century. The crucial documents signed at the summit were.

Rio Declaration (Earth Charter):

It is a document comprising of 27 principles, intended to generate the international cooperation and global partnership for the promotion of sustainable development. The 27 principles are as follows:

1. Human beings are at the centre of sustainable development. They are entitled to a healthy and productive life in harmony with nature.

2. States have the sovereign right to exploit their own resources with the responsibility to ensure the safety of their environment and of the areas beyond their national jurisdiction

3. The right to development must be fulfilled sustainably

4. Environmental protection shall constitute an integral part of the development process

5. To adopt policies for combating poverty

6. Environmentally vulnerable areas should be given priority in the field of environmental protection.

7. States shall co-operate to conserve, protect and restore the health and integrity of earths ecosystems

8. States should take up measures for changing consumption patterns and promotion of appropriate demographic policies

9. Adopt measures for exchange of scientific and technological knowledge

10. Handling of environmental issues by the participation of all citizens at the relevant levels

11. Implementation of effective environmental legislation

12. Adoption of a supportive and open economic system at international level

13. National laws regarding liability and compensation of pollution and other environmental damages to victims

14. Discouragement and prevention of the transfer of any hazardous substance or activity to other states

15. The precautionary approaches shall be widely adopted

16. National authorities should Endeavour to promote internalization of costs and the use of economic instruments

17. EIA as a national instrument shall be undertaken whenever required

18. Information about natural disasters and other emergencies shall be immediately given to other states

19. Information about the activities that may have trans-boundary environmental impacts shall be provided to the other states

20. Women should be involved to forge a global partnership

21. Involvement of youth to forge a global partnership

22. Involvement of locals and their traditional practices in the environmental management and development

23. Environment and natural resources of people under oppression, domination and occupation shall be protected

24. Respecting the international laws providing protection to the environment in times of armed conflicts

25. Peace, development and environmental protection are interdependent

26. Resolution of environmental disputes peacefully as per the UN charter

27. Co-operation between states and people in good faith and in spirit of partnership

Agenda-21:

It is not a legally binding document but a "work plan" or "agenda for action" or "action plan" for 21st century in all areas of environment and economic growth in sustainable way. It is a document which embraces the entire environmental and the developmental agenda. Agenda-21 has four main sections:

Section 1 (Social and Economic Dimensions):

Its main focus is the sustainable development of developing countries by better domestic policies and economic systems. It stresses on:

- Changing consumption pattern

- Better demographic dynamics

- Better human health and

- Better human settlements

It also included the need to integrate the environmental factors into laws, policy making, accounting and economic instruments.

Section 2 (Conservation and Management of resources for development):

This section promotes the understanding of integrated planning and management of various natural resources like:

- Land resources

- Mountains

- Water resources

- Agriculture

- Biological diversity and

- Forests etc.

It also focuses on the need for better information on risk assessment and management of toxic chemicals. Furthermore, it includes waste minimization, recycling, environmentally sound disposal of solid wastes, sewage and radioactive wastes.

Section 3 (Strengthening the Role of Major Groups):

This section includes entirely the statements on the importance of following nongovernmental sector in implementing sustainable development of Women, children, and youth, local authorities, business groups, trade unions, farmers, science and technologies.

Section 4 (Means of Implementation):

This section discusses the establishment of a sustainable Development commission; a new body under the Economic and social council, to coordinate the pursuit of sustainable development among international organizations and to monitor progress by the Non Governmental Organizations, governmental organizations and international organizations for the achievement of the goals of Agenda-21. It discusses the international legal instruments and mechanisms. It also discusses the issue of promoting public awareness of environmental issues through education and training, cooperation for capacity building in developing countries for implementing Agenda-21.

International Organisations

United Nations Developmental Programme (UNDP):

The United Nations Development programme was created in 1965 and has become one of the world's largest organizations of granting funds for economic and social development. Funds are provided to UNDP by the member states of United Nations. It is a committee to contribute to the United Nations Organization (UNO) to achieve its four basis aims of:

1. Perce and security

2. Humanitarian assistance

3. Development operations

4. Economic and social affairs

UNDP headquarters are located in New York and operations are managed through a network of 134 country offices. The UNDP administrator chairs the UNDG (United Nations Development Group) a committee of all the heads of United Nations funds. This committee promotes coherent United Nations action at country level through resident coordinator system and specific instruments such as the common country assessment (CCA) and the United Nations Development Assistance Framework (UNDAF)

Goals of UNDP:

1. To create enabling environment for sustainable development

2. To eradicate poverty

3. To achieve excellence in the management of UNDP operations

4. To achieve gender equality

5. To protect and regulate global environment and natural resource base for sustainable development

6. To reduce the impacts of emergencies, natural, environmental, technological and human disasters.

United Nations Environmental Programme (UNEP):

The United Nations conference on Human Environment (UNCHE) which was convened in June 1972 lead to the founding of UNEP (United Nations Environmental Programme), with its headquarter in Nairobi Kenya and responsible for coordination of intergovernmental measures for environmental monitoring and protection. UNEP has created worldwide awareness on emerging and existing environmental problems which lead to the upholding of different international conventions like:

- Montreal Protocol-1987

- Kyoto Protocol-1997

- UNFCCC-1992 and

- CITES-1993 etc.

It advocates a concept of environmentally sound development, which has lead to the adoption of the concept of sustainable development in the Brundtland Commission Report and the United Nations perspective document for the year 2000 and beyond Goals of UNEP: The tastes and goals of UNEP are varied and far reaching and include:

1. Developing national and international environmental instruments

2. Encouraging new partnerships in the private sector

3. Assessing environmental terms and conditions

4. Facilitating transfer of knowledge and technology

5. Promotion of environmental science and information

United Nations Conference on Trade and Development (UNCTAD):

It is a forum for trade and development and is committed to seriously improve and advance human development. It contributes to the international debate on globalization and the management of its consequences for developing countries. It also promotes policies at the national, regional and international level that are conducive to stable economic and sustainable development. UNCTAD has been designated as the task manager of the United Nations Inter-Agency task force on gender and trade.

Objectives of UNCTAD: The main objective of the conference on trade and development include:

i)   Poverty eradication

ii) Gender sensitive sustainable development

iii) Gender sensitive socio-economic development "by socio-economic impact assessment of trade related policies"

UNCTAD's Role in Trade and Development is:

i)   Research and analysis

ii) Consensus-building on policy issues

iii) Trade related technical assistance and capacity building for sustainable development.

iv) Sustainable and gender sensitive development.

Major Roles of UNCTAD:

1. Investment, technology and enterprise development: It includes

    • Analysis of foreign investment trends and their impact on development

    • Promotion of international investment and the understanding of issues involved

    • Use of new technologies in development

    • Devising of strategies for development of new small and medium sized enterprises

2. International trade and commodities: It includes

    • Promotion of development in developing countries through international trade

    • Their participation in trade related negotiations

3. The least developed countries: It includes

    • Organizing the three United Nations conferences on the least developed countries (Paris, 1981 and 1990; Brussels, 2001)

    • Special programme for least developed, land locked and island nations to promote their socioeconomic development through research, policy analysis and technical assistance particularly in capacity building.

World Health Organization (WHO):

The World Health Organization was established in 1946 under the United Nations programme of attaining highest possible level of health by everyone. Its role is to act as the directing and coordinating authority on international health work to establish and maintain objective collaboration with the United Nations, specialized agencies, governmental health administrations and professional groups.

Objectives of WHO:

1. To act as a coordinating authority on international health works

2. To assist government in strengthening health services

3. To establish technical and administrative services

4. To work for the eradication of epidemics

5. To promote cooperation among scientific and professional groups

6. To held conventions and make recommendations on various health matters

7. To conduct research in the field of health

8. To promote mother-child health care to foster the ability to live harmoniously in changing environments

9. To promote policies for maintaining environmental hygiene

10. To take all steps to attain the objectives of WHO

www.ingramcontent.com/pod-product-compliance
Lightning Source LLC
Chambersburg PA
CBHW082010190326
41458CB00010B/3143